A WORLD IN A SHELL

A WORLD IN A SHELL

Snail Stories for a Time of Extinctions

THOM VAN DOOREN

The MIT Press
Cambridge, Massachusetts
London, England

The MIT Press would like to thank the anonymous peer reviewers who provided comments on drafts of this book. The generous work of academic experts is essential for establishing the authority and quality of our publications. We acknowledge with gratitude the contributions of these otherwise uncredited readers.

This book was set in Adobe Garamond Pro and Berthold Akzidenz Grotesk by the MIT Press. Printed and bound in the United States of America.

Library of Congress Cataloging-in-Publication Data

Names: Van Dooren, Thom, 1980- author.
Title: A world in a shell : snail stories for a time of extinctions / Thom van Dooren.
Description: Cambridge, Massachusetts : The MIT Press, [2022] | Includes bibliographical references and index.
Identifiers: LCCN 2021035147 | ISBN 9780262047029 (hardcover)
Subjects: LCSH: Snails—Conservation. | Endangered species. | Extinction (Biology)
Classification: LCC QL430.4 .V36 2022 | DDC 594.3—dc23
LC record available at https://lccn.loc.gov/2021035147

10   9   8   7   6   5   4   3   2   1

For Emily

The stories told in this book have their home in the lands, seas, and skies of the Hawaiian Islands. I acknowledge the first peoples of these islands, the Kānaka Maoli, their ancestral connections to this place, and their ongoing efforts to aloha ʻāina.

Contents

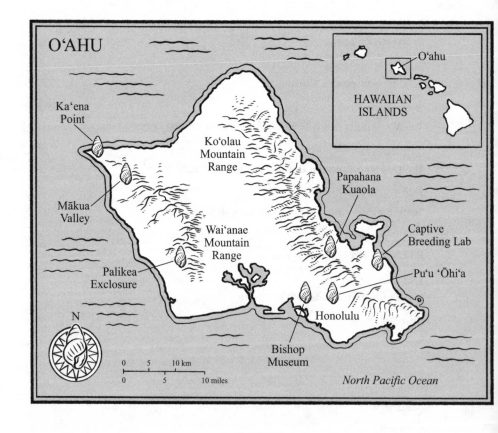

O'AHU

Ka'ena Point

Ko'olau Mountain Range

Papahana Kuaola

Mākua Valley

Wai'anae Mountain Range

Captive Breeding Lab

Palikea Exclosure

Pu'u 'Ōhi'a

N

Honolulu

Bishop Museum

O'ahu

HAWAIIAN ISLANDS

North Pacific Ocean

0   5   10 km
0   5   10 miles

INTRODUCTION
In Search of Snails

The dirt track we were walking on wove its way along a rocky ridgeline on the Hawaiian island of Oʻahu. Each side of the narrow path was lined by an assortment of windswept plants, foremost among them the hardy red-blossomed ʻōhiʻa. If we had walked this way 100 years ago, perhaps even 50, these branches would also have been laden with another colorful presence, that of the kāhuli, the tree snails. Upland forests like this one were once thick with their brightly colored forms; hundreds of snails might have been found living within a single tree. Unlike the leaf-eating garden species more familiar to most of us, these snails would not have harmed their botanical hosts. Rather, their diet consisted exclusively of the thin layer of fungi and other microbes that line the surface of leaves. As they made their nocturnal movements through the branches, they cleaned as they went.

But none of these snails could be found here anymore. We walked through a landscape now missing most of the diverse species that once called this place home. But we walked toward snails, nonetheless.

My guides were two conservationists, Dave Sischo and Kupaʻa Hee. More accurately, I was tagging along on one of their routine trips to one of the few places where snails can still reliably be found in the Waiʻanae mountain range; indeed, one of the few places where they can now be found in abundance anywhere in the Hawaiian Islands. After about an hour of walking up and down, through the undulating and often muddy terrain, we arrived at our destination. The forest opened into a clearing, and in front of us stood a shoulder-height green metal fence: the Palikea exclosure.

Encircling roughly 1,600 square meters of vegetated land, the exclosure is a place of refuge for rare snails. It is called an "exclosure" rather than an "enclosure" because its function is not primarily to keep these snails in, but rather to keep others out. Specifically, the fence aims to exclude the many predators of snails that have arrived in these islands with their human inhabitants, predators such as rats, Jackson's chameleons, and most important of all, the rosy wolfsnail (*Euglandina rosea*), a carnivorous snail that tracks and consumes the local species with devastating efficiency.

This exclosure is a key component of the work of the Snail Extinction Prevention Program (SEPP), a partnership between the Hawai'i State Government and the US Fish and Wildlife Service that is headed up by Dave. In all there are now nine such exclosures across the island of O'ahu, mostly concentrated in the Wai'anae Range, with two more in the planning and construction stages. In addition, exclosures are now being established on some of the other islands, with two completed on Lāna'i and one on Maui (and two more in development). Many of these exclosures were constructed by and are managed in collaboration with the US Army, as part of their legal obligation to offset the ongoing impacts of military operations on the environment. In addition to these fenced areas, SEPP also maintains captive populations of a range of endangered snail species in its laboratory, a large trailer fitted out with refrigerator-like "environmental chambers" that function as an ark of sorts for their slimy inhabitants. The sad reality, however, is that even with these various facilities, SEPP is able to protect only a fraction of Hawai'i's threatened snails, and with each passing year more and more species find themselves in need.

To put the matter simply, the snail situation in Hawai'i is dire. In our conversations, Dave and his colleagues described a rapidly unfolding crisis on an immense scale. These islands were once home to one of the most diverse assemblages of snails found anywhere on earth; hundred and hundreds of unique species of diverse shapes, colors, and sizes. Almost two-thirds of these species are thought to already be extinct, most in the past 100 years or so. Equally as concerning, the majority of those species that do remain are headed quickly in the same direction. Dave estimates that as many as 50 species are currently right on the brink; unless something dramatic changes in the next few years to improve their situation, they will be gone—or at best existing only in one of these captive spaces.

In the face of this escalating crisis, the SEPP team is engaged in what Dave describes as an "evacuation," trying desperately to locate the last remnants of snail species and get them out of harm's way. In the rugged terrain that makes up much of the main Hawaiian Islands, this work frequently involves long hikes, and when hiking isn't possible, the use of helicopters to access remote areas. In the past, conservationists took only a small number of snails from these populations as a backup, leaving the rest in the forest. Over the last few years, though, things have deteriorated rapidly. Dave explained that in this time he has seen about 15 formerly robust populations, each comprised of hundreds of snails, completely disappear. Some were the last known free-living populations of their kind, like that of the tree snail *Achatinella lila*, a species with varied shells of yellow, green, and the deepest mahogany brown. As a result, despite initial reluctance, the SEPP team are now often pulling out every snail they can find, carefully packing them into containers, and bringing them into the lab or an exclosure: "We're just getting them the hell out of there before it's too late."

The causes of this decline are complex and multifaceted. The most significant threat faced by Hawai'i's snails today comes in the form of the predators—chameleons, rats, and carnivorous snails—that the forest exclosures have been designed to protect them from. As Dave explained to me, the rosy wolfsnail is the single largest threat many of the local snail species face today: "They're our worst predator and unfortunately, they've made it up into the highest forest reserves where some of the last refuges are for these snails. The rosy wolfsnails are vacuuming up everything."

These impacts come in the wake of a longer history of snail decline. Key to this story is the widespread loss of the islands' forests. To a certain extent, this process began with the arrival of Polynesian peoples roughly 1,000 years ago.[1] As they cleared forests to make room for kalo (taro), 'uala (sweet potato), and the other agricultural plants that would allow these islands to sustain human life, they removed and transformed snail habitat. At the same time, the rats that they brought with them—voracious consumers of seed and fruit—are thought to have begun a process of significantly altering the composition of the islands' forests.

But these impacts were drastically scaled up after the arrival of European explorers in the late eighteenth century. As the animals such as cows and goats that they introduced and helped to establish spread out across the islands, they significantly damaged or removed the understory of vast areas of forest. From the mid nineteenth century, Europeans and Americans who had settled permanently by this time drove a process of widespread land transformation in which areas of forest were cleared to make way for ranching and plantation agriculture, and then from the twentieth century for tourist, urban, and military developments.

On Oʻahu, where most of the stories in this book take place, the situation is particularly bad. As the biologists Sam ʻOhu Gon and Kawika Winter have summed it up: "fully 85 percent of the land that once was home to a diverse endemic set of ecosystems and species had been entirely lost, leaving small remnants only on the highest summit crests of the island."[2]

The arrival of Westerners also kicked off a period of intense shell collecting. Beginning in the 1820s, but with a real fervor during the final half of that century, people gathered up these shells on a staggering scale. Millions of live snails were reduced to specimens in huge private collections, many of which were well in excess of 10,000 shells. This collecting craze was further fueled by interest from beyond Hawaiʻi's shores, as museums and naturalists the world over sought to acquire their own specimens.

Today, the impacts of climate change are being added to this difficult situation, transforming rainfall and temperature patterns across the islands in ways that can't yet really be understood. Modeling, however, indicates that we can expect generally hotter and drier conditions, which do not bode well for moisture-dependent snails.

Within the protected walls of the Palikea exclosure that day, however, the snails were doing well. Small groups of *Achatinella mustelina*, a striking white and brown tree snail, were gathered together on tree trunks and branches; other incredibly rare ground-dwelling snails were making their way through the decaying leaves, like the beautiful *Laminella sanguinea* with its red and orange shell marked with a series of distinctive lightning-bolt stripes that run from apex to aperture. The fragile bubble of life within this enclosure has been carefully created and tended, and is now home to four critically imperiled snail species. But outside these walls, things continue to worsen. Recovery or restoration of many of these species is simply not possible, at least for the time being. Instead, Dave explained, "for the next decade, I foresee that our role as a program is just preventing extinction . . . We're essentially manning the lifeboats."

A few days later I found myself at the Bernice Pauahi Bishop Museum in Honolulu, among the cabinets that house their incredible collection of snail shells. Each drawer that we opened revealed another set of wonders, another surprising color or variation in shape or size. In one drawer we encountered the delicate, translucent forms of *Succinea lumbalis*, in another the gray conical shells of *Newcombia cumingi*, in other drawers, the tiny zebra-striped shells of *Laminella aspera*, not more than a few millimeters in length when fully grown and yet intricately patterned, nonetheless. In many other drawers, we found the colorful shells of the larger *Achatinella* tree snails, some with bands and stripes, others sporting designs reminiscent of tweed or tortoise shell. Drawer after drawer, cabinet after cabinet, row after row, we moved through the museum's malacology collection, ultimately seeing only a tiny selection of their shells.

My guide was Nori Yeung, the curator of this repository of snail shells from all over the Hawaiian Islands and the broader Pacific region. I had asked Nori to show me the collection and speak with me about its history and significance. In no small part, I just wanted to try to wrap my head around the diversity of snails once found in these islands. The sad reality is that with so many of these species already gone, or exceedingly rare, the museum is now the only place that you can go to get a sense of the richness of Hawai'i's snails.

But cabinets of empty shells cannot really capture that diversity. A large, brightly colored, snail moving through damp leaves in the forest, its shell glistening as it goes, is a sight to behold. After death, the shells that snails leave behind quickly lose their luster, and then their color, much of which resides in a thin, living layer on their surface: the periostracum. Many of the shells in the museum are still remarkably beautiful, but a visit to these cabinets requires a great deal of imagination, more and more as the years go by, to think these creatures back into their worlds.

Among all the shells I encountered that day, those of *Carelia turricula* stood out. This species of ground-dwelling snail is thought to have been the largest found in all of the Hawaiian Islands. Their tall conical shells in varied shades of purple and brown filled several drawers. Some of the adult shells I saw were about 5 centimeters in length, but others about half that size again have been reported. These snails were once found widely on the island of Kaua'i. Writing in 1887, the naturalist David D. Baldwin reported discovering alluvial deposits that contained multitudes of their shells and surmised that they must, at some point in the not too distant past, have been "very abundant."[3] But even in his day they were rare; today, shells like these in museum drawers are all that remains of the species.

As I looked down on the shells of *Carelia turricula* that day, they seemed so improbably large and awkward. I tried to imagine a snail moving along the ground with a narrow 8-centimeter protuberance sticking out behind it. Then I tried to imagine a landscape thick with these remarkable creatures. I asked Nori about them, and she replied with the same sense of speculative wonder: "Just to see them alive and crawling would have been so amazing." But try as I might, I can't quite imagine that sight.

It is not well known that, among its many biological riches, Hawai'i is a land of snails. While snails can be found all over the world—indeed, they inhabit every continent and island archipelago outside the Antarctic—very few other places supported anything like the variety found in this island chain. To date, over 750 species of Hawaiian land snails have been recognized by taxonomists. The actual number though is thought to be considerably higher; perhaps somewhere in the region of 1,000. But even at around 750, the tiny patches of land that form these islands were once home to roughly two-thirds the number of snail species that can be found in the whole of continental North America, a land mass about 1,700 times the size. What's more, almost all of these species—over 99%—were endemic to these islands,

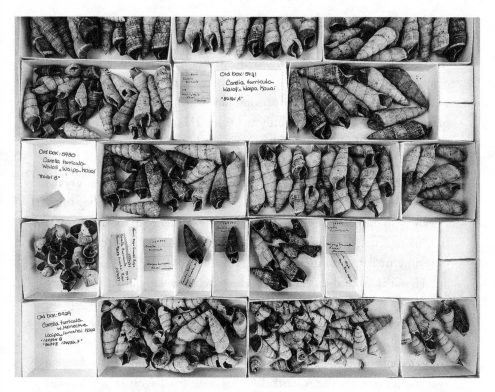

Carelia turricula shells in the Bishop Museum collection.

found here and nowhere else.[4] By any reckoning, Hawai'i is a remarkable place when it comes to snails.

All of these species can be divided, somewhat crudely, into two groups: the tree snails and the ground snails, distinguished by their primary habitat and diet. These snails vary greatly in size, color, and shape, but not one of them leads the kind of leaf-munching life that most people associate with their kind, eating living vegetation like the tender lettuce plants in your garden. In other parts of the world there are also carnivorous snails that focus on consuming other animals; these too were not to be found among the islands' snail fauna until they were introduced by people. Instead, in Hawai'i the tree snails live among the branches and consume the thin layer of fungi and other microorganisms that line the surface of leaves, using their sandpaper-like radula (effectively, a tongue lined with hundreds of tiny teeth) to scrape harvested materials into their mouths. The ground snails, in contrast, spend their lives down in the leaf litter, and use their own specialized radulas to eat dead and decaying plant matter, recycling it back into the soil.

Among the tree snails, many people have long been captivated by those of the genus *Achatinella*, found only on Oʻahu. These snails, with shells about 2 centimeters in length when fully grown, could once be found in a variety of colors and patterns: greens, yellows, and reds, many with spirals, stripes, swirls, and other patterns. Historical records from the nineteenth century tell us that these snails were once not just common, but abundant. They hung thickly from the trees like clusters of grapes; they glistened in the damp forest like "living jewels."[5] The genus *Achatinella* once included 41 species. Today, only nine are left, all of them critically endangered.

Hawaiʻi's ground-dwelling snails have been similarly impacted. Many of these species are tiny brown creatures, barely visible on the forest floor. Others are brightly colored and striking to the eye, like *Laminella sanguinea*, with its lightning-bolt stripes, or *Amastra spirizona*, with its beautiful brown conical shell encircled by a pale, spiraling band. The family to which these two species belong, the Amastridae, was once the largest in the islands, including at least 325 species; a mere 23 are thought to remain.

The staggering rate of extinction among Hawaiʻi's snails is part of a larger global trend. We are living in the midst of the earth's sixth mass extinction event. All over the world species are disappearing en masse. Of course, extinction is an unavoidable fact of life. Given time, all species eventually go the way of the dinosaurs. But in our time these losses are mounting up: species are thought to be blinking out somewhere between 100 and 1,000 times faster than the normal rate.[6]

In the midst of all this loss, snails have been particularly hard hit. This is a fact that is generally not well known, even among conservationists. Snails, it seems, do not always generate a lot of concern. However, according to the International Union for Conservation of Nature (IUCN), whose job it is to keep the official lists of these things, worldwide there have been more documented extinctions of Gastropoda—that is, snails and slugs—than there have been of all the birds and mammals combined.[7] While places like

Hawai'i are concentrated epicenters of gastropod extinction, and indeed many other islands have also experienced high rates of loss, this is a world-wide phenomenon. Robert Cowie, a biologist at the University of Hawai'i who has written extensively on this decline, summed the situation up simply: "both in terms of total numbers and the proportion of their global diversity, gastropods are one of the hardest hit of the major taxonomic groups."

The term "Gastropoda" was coined by the naturalist Georges Cuvier in 1797, Gastéropodes in his native French. Conjoining the Greek terms for stomach and foot, the name provides a simple reference point for these creatures that walk on their bellies.[8] There is something strangely fitting about the fact that it was Cuvier who gave us this nomenclature. After all, it was he who also introduced modern science, and the Western world, to the notion that species might become extinct.[9] Just over 200 years later, the fact of extinction has become all too real; and it now turns out that another of Cuvier's contributions to science—the Gastropoda—has become one of the most significant embodiments of this process of loss.

In fact, the gastropod situation is actually much worse than the official records might lead us to believe. All such lists are able to cover only a fraction of the earth's diversity. We just do not have the necessary data to know the status of each individual species. In fact, best estimates indicate that science has only identified and formally *described* (not assessed) a fraction of the Earth's species, perhaps somewhere in the order of 20%.[10] While birds and mammals have for the most part been pretty well taken care of, invertebrates such as insects, snails, and spiders haven't received anywhere near the same amount of attention. What's more, there are a great deal more of them; invertebrates are thought to make up roughly 99% of the animal kingdom.

So, while the IUCN's *Red List* includes only about 900 documented cases of extinction among all of the various kinds of animals over the past 500 years, the real figures are undoubtedly much higher. Again, it's likely that most of the extinctions of larger mammals and birds have been accounted for in this list, but the vast majority of those that have occurred among the invertebrates have not. Their sheer numbers, combined with the uneven-ness of research and the rigorous evidential requirements of the IUCN, mean that most *known* invertebrate species end up being classified as "data deficient"—and this is after excluding the many more invertebrate species that we haven't yet even discovered, let alone formally assessed.

As a result, the IUCN officially records only 267 gastropod extinctions globally in the past 500 years. Clearly, a great many are missing: all experts agree that in Hawai'i alone there have been more snail extinctions than this. In one recent study, Cowie and his colleagues offered what they called "a more realistic, albeit less rigorous, assessment" of gastropod extinctions around the world. They found almost 1,000 species of snails that are known or highly likely to be extinct. The researchers noted, however, that their findings are significantly biased by the fact that they can include only species for which information is available. Their best estimate is that the actual number of gastropod extinctions during this period is between 3,000 and 5,100.[11]

Even in the context of this widespread global snail loss, the situation in Hawai'i stands out. These were islands once particularly rich in snail life, and they have been among the worst affected by this decline. Nori and her partner Ken Hayes, also based at the Bishop Museum, have done the most comprehensive survey of the islands' snails. Working in dense, often inaccessible, forests, it is incredibly difficult to make definitive statements about the presence or absence of tiny snails, some of them only a few millimeters in size as adults. Nonetheless, drawing on extensive historical survey materials held in museums and thousands of field hours at sites throughout the island chain, they have estimated that roughly 450 species have already been lost.[12] Most of these losses have taken place in the past 100 years or so, with numerous species that were found during field surveys in the early 1930s now having entirely disappeared.[13]

Equally as concerning, Nori and Ken's preliminary surveys indicate that the remaining 300 species are in a great deal of trouble. Of them, over 100 are "critically imperiled"—reduced to a single known population in the wild. A further 120 species are down to just two or three populations. Their research shows that a grand total of 11 species can be categorized as "stable."[14] With these numbers in mind, it does not seem unreasonable to suggest that Hawai'i's snails are undergoing their own, private, mass extinction event.

And yet, despite this incredible and ongoing loss, a mere 12 of the 300 or so remaining species are officially listed as threatened and protected under the federal Endangered Species Act (ESA). All of those that have made it onto this list are among the more conspicuous, large, and brightly colored tree snails.[15] For the rest, as is so often the case for invertebrates around the world, there simply has not been the kind of long-term, intensive study that is needed to qualify for listing.

While funding for snail conservation in Hawai'i has slowly grown over the past few decades, as Dave put it, "the amount of money that we need to do what we need to do is just not there." Conservation work is underfunded all over the United States, and indeed around the world, but Hawai'i's snails face the combined challenge of both geographical and taxonomic biases working against them. Nearly one-third of all of the listed endangered species in the United States are found only in the Hawaiian Islands. And yet Hawai'i receives less than 10% of the federal funding allocated specifically to endangered species.[16] At the same time, invertebrates in general are relatively underfunded. As Nori and Ken pointed out in a recent paper, while listed vertebrate species like mammals and birds received an average of $2 million of federal funding each in 2015, the invertebrates each received about 6% of that figure.[17] In other words, endangered invertebrates in Hawai'i are in a particularly tough spot.

Despite these considerable challenges, something must be done. It is hard to overstate the significance of our present moment: to put it simply, our current decade will decide everything for Hawai'i's snails. While numerous species have already been lost, many of those that remain have been left so diminished that they are now beginning to blink out, one after the other, in quick succession. But at least some of them can still be held onto. Doing so, however, will require a concerted effort; it will mean learning to see and value snails in new ways.

With the exception of the odd outlier, like the beautiful butterflies and the brainy octopus, invertebrates have an image problem. Sadly, it isn't one that they, or we, can readily ignore as it carries significant consequences. The biologist Timothy New has described a "crisis in invertebrate conservation" that has its roots in no small part in this generally poor popular opinion, a "public prejudice against invertebrates."[18] As a result, the loss of countless species that crawl, creep, buzz, and flutter is largely ignored, and occasionally even celebrated, in public discussions of biodiversity loss. It is interesting to note, for example, that the first iteration of the Endangered Species Act didn't even include invertebrates (or plants, for that matter). So, while the decline of charismatic fauna like elephants, tigers, and whales frequently makes headlines—which is not to say that anywhere near enough is being done for these species—all around us a quieter systemic process of loss is relentlessly ticking on, with many invertebrates slipping out of the world entirely unnoticed.

I must admit that I hadn't given snails, or most other invertebrates, a great deal of thought until I began the research for this book. But over the last several years, I have become convinced that if we take the time to really look at them, we cannot help but discover that they are remarkable beings. They are creatures with their own unique stories, relationships, and significances; species that are entitled not to be blotted out simply because it is easier or more profitable for us not to care. Beyond their own intrinsic value, many of these invertebrates are also at the core of vital ecological functions that we, and the wider world, are utterly dependent on. They are the decomposers, the pollinators, the seed dispersers, the nutrient cyclers, and more. In biologist E. O. Wilson's memorable phrase, they are "the little things that run the world."[19]

While working on this book I have had numerous conversations with people who have been surprised, and often a little bemused, by the focus. Many have asked me why I am spending my time in this way. Surely there are more important conservation issues to be concerned with? The firm conviction that guides this book is that learning to see and appreciate invertebrates is a vital task for our time, a crucial but much neglected aspect of responding well to our current period of mass extinction. We need to move beyond caring for only those species that catch our eye, are immediately relatable, or are charismatic. We need to cultivate more expansive and inquisitive modes of appreciation that draw us out into the vast and tangled webs of living beings of all kinds that together constitute our world. Snails have provided a silvery pathway into these issues for me. My hope is that through this book they might do so for others, too.

The rain poured down, drumming loudly on the tin roof above us. I was sitting inside a small pergola with Cody Pueo Pata at Papahana Kuaola, an educational facility on the windward side of O'ahu where the teaching is grounded in Hawaiian skills and knowledge. Pueo is a kumu hula, a teacher of the art of hula, which he practices in a traditional form that is rooted in a deep study of and respect for the 'āina (the land) and the many deities that reside within it.

I had asked Pueo to meet with me to talk about snails and the varied ways in which they have woven themselves into the lives and culture of these islands' first peoples, the Kānaka Maoli. Pueo began by telling me that in the traditional stories, snails "often accompany our forest goddesses." He went on to explain that "they're believed to produce chirping sounds. So, when these goddesses appear, they're surrounded by birds and the chirping kāhuli, or the different kinds of pūpū, or shells, that grow in the forest." But, he asked: "When those things are no longer in the environment, does that mean the akua [deity] is no longer there?" With the loss of so many snails, Pueo wondered if the forest might be becoming a less appropriate place for the gods to reside within.

In posing this difficult question, Pueo referenced what is, without doubt, one of the most prominent themes associated with snails in Hawaiian mo'olelo and mele (stories and chants): the notion that they sing in the forest. In fact, alongside the Hawaiian name kāhuli, another common name for land snails is pūpū-kani-oe, literally translating as "shell sounding long." But in the stories told about them, the snails are not said to sing at any old time. Rather, their singing is deeply meaningful, often said to occur as a sign that after a series of adventures, changes, or turbulence, all is pono again—all is righteous, correct, and good.[20]

A World in a Shell offers a set of snail stories for this time of extinctions. This book is grounded in the understanding that storytelling about extinction and biodiversity loss is a vital task. Stories can thicken and enliven our understanding of what particular extinctions mean and why they matter; they can allow us to acknowledge and even mourn; they can also be transformative, drawing us into new worlds, into appreciation, complexity, and responsibility.[21]

In telling snail stories, this book aims to cultivate an appreciation for these animals and the significance of their loss: to draw us into their remarkable miniature worlds, and then out beyond them into an expansive engagement with the many ways in which snails craft and share these worlds with others. This book is about snails' modes of perceiving and interpreting the world, from their slime-centered navigation to their social and reproductive proclivities; the immense journeys that brought them across oceans to these islands; the histories and ongoing practices of learning and knowledge-making about our world that they have been part of; their intimate relationships with Kānaka Maoli as expressed in chants, songs, and stories, but also in ongoing struggles for land and culture. In short, it is a book about the world of possibilities and relationships that lies coiled within each of their tiny shells.

In this way, while this is a book that is attuned to the particularities of these islands and their snails, it is also one that is concerned with the far larger problem of biodiversity loss that we are facing all over the world today. At the heart of this book is a simple question: How might paying attention to snails in this particular time and place help us to rethink and engage differently with our escalating extinction crisis?

In exploring this question, this book draws on research in the natural sciences and the humanities. There is an important difference here from many other books about extinction for a general readership, which tend to be written by biologists or by journalists who have spent a lot of time talking to biologists. This is not unreasonable; after all, these scientists are the obvious experts on the topic. But while the insights of biologists are a vital part of the picture—and I have drawn on them in great depth in this book—they are not the only stories to be told. What tends to be left out of these accounts are all of the other complex human relationships at work and at stake in extinction. Delving into these issues requires also talking to other kinds of people, as well as engaging with diverse cultural, historical, and philosophical questions.

I am an environmental philosopher by training, but from the beginning of my career I have been passionate about field research: interviewing local people and traveling with them to see and to learn. In company with a small but growing group of other scholars, I now think of myself as a field philosopher.[22] For roughly the past 15 years, my research and writing have focused on extinction, working to better understand why the disappearance of so many forms of plant and animal life at our present time matters so much, and to tell stories that bring us into an encounter with those complex processes of loss.

Research for this book has taken me into the forests and the mountains in search of a wonderful variety of snails; to laboratories that are working to safeguard and breed them in captivity as well as to better understand their lives and needs; to museums to see the shells and historical records that have simultaneously shaped our understanding of the past while guiding conservation action today; along rocky coastlines in search of ancient shells, archives of past lives and environments; and even into a military live-fire range and weapons disposal site where we needed to keep a constant lookout for unexploded ordnance. More than anything, though, I have spent time listening and talking with a range of people, in their living rooms and offices, learning about what Hawai'i's snails and their disappearance mean for them.

I am firmly of the view that there is no singular extinction phenomenon. Each species leaves this world in its own particular way, having its own unique impacts on the contours of local lives and landscapes. There are no simple answers to our current period of mass species extinctions. Instead, what this time calls for is a multiplication of stories, an explosion of voices seeking to share, to challenge, to understand, to bear witness, and ultimately to become responsible for the ongoing loss that is unfolding all around us.

One crucial part of this effort is learning to see and narrate extinction as a process that is much more than an environmental concern that takes place 'out there' in nature, somehow separated off from human life. Extinction is a profoundly human process. But extinction also implicates and involves us all differently. The stories I tell in this book emphasize the way in which the loss of Hawai'i's snails is tangled up with larger processes of globalization, colonization, militarization, climate change, and more.

In this way, while taking up elements of popular science and nature writing, this book aims to play a role in challenging and stretching these genres. For the most part, humans tend to appear in one of two main guises in popular writing about extinction: either they take the form of an individual conservationist or a group of such individuals, struggling, often heroically, against the loss of species, or they are present in the form of an amorphous, threatening, "humanity" whose actions are in one way or another causing these losses. But if we look a bit closer, things are always more complex on the ground. Missing from such stories are the diverse and unequal ways in which different communities are exposed to and suffer through extinction, as well as the very particular systems of political, economic, and cultural life that are responsible for any given extinction. Paying attention to these dynamics highlights how thoroughly inseparable processes of environmental destruction are from specific historical and ongoing dynamics of violence and dispossession.[23]

In recent years, the notion that we have entered a new geological epoch—the Anthropocene—has risen to global prominence, being taken up everywhere from scientific discussions to gallery exhibitions. According to this view, human activity is now playing an increasingly significant role in shaping environments all over the planet, including impacts on climate, biodiversity, the nitrogen cycle, and much more. On one level this framing is helpful, drawing attention to the scale of contemporary destruction. But at the same time, as numerous scholars have noted, "Anthropocene talk" risks covering over significant differences between human communities and the particular ways in which they are responsible for and exposed to escalating environmental change.[24] Attending to this kind of difference means moving beyond stories about "humanity" and "anthropogenic extinction."

In exploring the main factors driving the disappearance of snails in Hawai'i, we see that it is particular forms of human life that are responsible, in most cases processes of extraction, appropriation, and wealth generation.[25]

The vast majority of the snail habitat cleared in the nineteenth and twentieth centuries was lost to large-scale sugar and pineapple plantations or to cattle ranches, much of it having been earlier logged for sandalwood and other valuable timbers that were exported to global markets. It was efforts to protect agricultural crops, albeit poorly planned and executed ones, that led to the introduction of the rosy wolfsnail in the 1950s as a biocontrol agent for the giant African snail, introduced earlier.[26] Along the way, right up to the present day, the US military has also had significant impacts on snails and their habitat in the large areas of often highly biodiverse land that they routinely blow up for training purposes across these islands.

These stories are not about 'humanity'; they are tales of particular people, places, and processes. Getting at this complexity matters for understanding how and why extinctions happen.

Many of these same processes have impacted significantly on Kānaka Maoli. The ranches and military bases that took snail habitat also swallowed up their traditional lands. The collecting and documenting of snail life in the nineteenth century were, in their own ways, part of the larger processes of settlement and appropriation through which the Hawaiian monarchy was overthrown, and the islands became a US territory and then a state. In Hawai'i, this unraveling of Indigenous lifeworlds has been thought about by Kanaka scholar Jonathan Kay Kamakawiwoʻole Osorio as a process in which, despite concerted opposition from local people, "colonialism literally and figuratively dismembered the lāhui (the people) from their traditions, their lands, and ultimately their government."[27]

The story of the decline of Hawai'i's snails cannot be told well outside this larger context. What does this decline mean to people who have inherited stories of singing snails? What becomes of these stories, and the knowledge and meaning that they carry, when there are no more snails left in the forest? How are Kanaka connections to land and culture further challenged? The destruction of the worlds of snails and Kānaka Maoli are inextricably entangled.[28] But so too are processes of resistance, of conservation, and of crafting future possibilities for shared life. As we will see, in at least a few places in Hawai'i and the broader Pacific, Indigenous peoples and snails are engaged in vital practices of solidarity—with each other and with scientists, lawyers, and other concerned people—to protect their lands against the destructions wrought by the military and others.

The snail stories told in this book try, in their own small way, to acknowledge these histories, presents, and futures. Moʻolelo (story) is central to Hawaiian life and culture. As Ty P. Kāwika Tengan has noted, Kānaka Maoli "have always made and remade their identities through the re-membering and retelling of moʻolelo, especially in times of rapid change that threaten their continued existence as a people."[29] This is a work that reaches into the past, but that is ongoing and dynamic, one that seeks to hold open, generate, and protect possibilities for what Noelani Goodyear-Kaʻōpua has described as multiplicitous, Indigenous futures.[30]

In diverse and unequal ways, we are all at stake in extinction. It threatens the ecosystems that sustain us, the cultures and systems of both meaning and mystery that animate our lives, and, in the indifference of so many of our responses to it, extinction also wounds and threatens our humanity. As extinction remakes lives, landscapes, and possibilities, it forces us to ask: Who are we and whom might we become when species *disappear*?[31]

If life is a tapestry, what does it mean to un-weave it, to pull out one strand, or, in this case, a whole handful of strands? What else comes undone when this happens? Ultimately, this book is a close exploration of these processes of unraveling, of the causes, consequences, and meanings of an ongoing extinction event. It shows that the absence created by extinction is not a nothingness; rather, it is an ongoing unravelling that ripples out into the world in myriad ways. Viewed in this way, extinction is not a short, sharp event but a drawn-out *process*: a space in which many living beings, human and not, must make their lives. Good stories help us to see and become responsible for these rippling processes. But they also create openings into something else, into new forms of resistance and recuperation.

From this perspective it is clear that while conservation efforts like the ones detailed in this book are vital, they are also not enough. Broader changes are needed.[32] In order to achieve them, however, I sincerely believe that we require a fuller appreciation of the diverse forms of plant and animal life that are disappearing around us, as well as just how much is at stake in their loss. It is this kind of appreciation that the stories in this book seek to cultivate.

In taking up this story work, this book is an effort to "pay heed to the voice of the kāhuli," as the renowned scholar of Hawaiian language, Puakea Nogelmeier, put it to me, referencing a traditional chant. It is an effort to slow down, to listen, and to share stories that might play some small role in the crafting of better possibilities for life.

I looked down at a single snail, alone in a plastic case. Around its tiny form, a collection of leaves and small branches from native plants had been carefully assembled to provide both food and some semblance of the trees its ancestors and relatives once lived among. But none of those other snails could be found here now. In fact, they were not to be found anywhere. After more than a decade of searching in the forests of the Koʻolau Range on Oʻahu, researchers were now confident that this singular individual was all that was left of its species. Born into captivity in this very room almost a decade earlier, it had become the living embodiment of its kind. Millions of years of evolutionary history condensed into its fragile form: countless snail journeys through the branches in search of food, shelter, or company, generation after generation of beautiful, slimy life in the trees. All of this would come to an end with the death of George.

Around us stood six environmental chambers that looked quite a lot like old refrigerators, humming noisily in the background. As their name implies, the function of these chambers is to reproduce the environmental conditions—principally, the temperature and humidity—appropriate for their particular inhabitants. Each of them was filled to the brim with an assortment of plastic containers, most of them home to multiple snails of a single species, both adults and their young. Some of these populations were thriving, reproducing at rates that could barely be contained; but many others were in significant decline, heading swiftly in the same direction as *Achatinella apexfulva*, George's species.

This was my first encounter with Hawaiʻi's snails. It took place in 2013, a few years before I started work on this book. The visit left a lasting impression on me, and in many ways this book is a product of that encounter with George. Our meeting took place in a small room on the campus of

George, the last snail of the species *Achatinella apexfulva*, photographed by the author in 2013.

the University of Hawai'i at Mānoa in Honolulu. This was a couple of years before the establishment of SEPP, when the only captive breeding facility for snails was this space that had been set up on a shoestring budget in the mid-1980s by Mike Hadfield. I was in Hawai'i studying some of the islands' many threatened bird species and had been put in touch with Mike by a mutual friend, Donna Haraway. Mike offered to show me the snail lab, and I jumped at the opportunity.

Over the years since I started work on this book, I've been able to spend many more hours and days in the company of snails, as well as with some of the dedicated people who care passionately about them and their future. I encountered George a few more times during this period. Each time I did, I experienced the same feelings of profound sadness, of hopelessness, that I suspect accompany most encounters with an "endling," the term used to refer to the last individual of a kind.

Then, in the early hours of the morning on January 1, 2019, George died. With this little snail went the species *Achatinella apexfulva*. This species was once widely distributed across the central-northern Koʻolau Range. It was commonly collected in the nineteenth century and was the first Hawaiian land snail to be afforded a Linnaean name, *apexfulva* referring to the characteristic yellow-tipped shells of many adults. It was also the first species to be taken to Europe, in the form of a traditional Hawaiian lei.[33]

By the time George was born, however, the species was exceedingly rare. In 1997, 10 of these snails had been brought into the lab. At that time *A. apexfulva* had been recently rediscovered, having been thought to be among the roughly three-quarters of the genus already lost. A few young—usually referred to by scientists as keiki, using the Hawaiian term for a child—were born to these captives. George was one of them. In the years around 2010, however, a wave of death spread through the lab, thought to have been caused by a pathogen. All of the other snails of the species, adult and keiki alike, were killed. George spent roughly the last six years of his life in isolation, first in the facility at the University of Hawaiʻi where we met, and then at the successor lab established by SEPP in 2016.

In death, George immediately became a minor celebrity. As the first extinction of 2019, he made headlines across the United States and to a limited extent around the world, including in the *Guardian*, the *New York Times*, *Scientific American*, and the *Huffington Post*. He even prompted a beautiful feature essay in the *Atlantic*. (George was generally referred to as a "him" in this coverage, even though, like most land snails around the world, George was a hermaphrodite.)

In life, George had often attracted attention, prompting various articles and the melancholic interest of visitors like myself. As Dave, who spent many years looking after George in the SEPP facility, put it in his obituary: George was "an ambassador for the plight of the Hawaiian land snails." But in death, this little snail's celebrity status skyrocketed, at least for a short time. The attention stood in stark contrast to the general lack of popular interest in the ongoing and escalating loss of snails in the islands. Dave admitted to finding all the coverage of George's passing a little shocking. In an email, Mike asked me: "Who would ever have expected this from the death of a snail?"

George's death reinforced for me the fact that endlings can be highly charismatic figures—even, it seems, if they're snails. The nature of this charisma is a bit more difficult to pinpoint, though. Of course, George's story is immediately compelling and unsettling: his lonely decade in captivity grabs people. His death seems to make extinction relatable, locatable, and narratable.[34] But there is also something a bit disturbing about this celebrity.

Hawai'i's land snails have been disappearing en masse for well over a century. And we've known that they're disappearing: it has been documented and even discussed in newspaper articles since at least the 1870s. What's more, George was the last of his kind for over a decade, quietly living out his life in captivity. But during none of these earlier times did we see anything like the kind of interest that emerged after George's death.

While this much-needed attention on the plight of Hawai'i's snails has the potential to do some good, it was ultimately short lived. Roughly 6 months later, it had completely disappeared. Even George's story was ultimately able to summon up only a fleeting concern, centered on a moment of spectacular tragedy. Perhaps, in our media-saturated world, this is the best we can hope for?

While most of this book was already written when George died, his death prompted me to think again about its purpose. More than anything, we need to learn to tell more complex and sustained stories about extinction. My point is not that we ought to ignore George and other endlings; we owe them much more than this. Rather, my contention is that endling stories are not on their own adequate. Other modes of communication are needed, modes that don't rely solely on the charisma of the last individual to communicate what is at stake. Every snail—each individual and each species—is remarkable in its own way if we take the time to learn to see them. The tragedy of extinction cannot be distilled into a singular death. It is a much more complex and drawn-out process of relationships, possibilities, and worlds, unraveled and redone. Guided but not contained by the tragedy of George's life and death, this book is an effort to tell these kinds of snail stories, both of and against extinction.

1 THE WANDERERS
A Foray into Worlds Written in Slime

As I stepped down off the ladder into the exclosure, I must admit to being just a little bit disappointed. I had imagined, very naively it now turned out, that this place would be crowded with brightly colored snails. Over the previous hour or so that it had taken us to hike to the Palikea exclosure, my anticipation had been slowly building. I had joined one of SEPP's routine monitoring and maintenance trips, traveling with Dave Sischo and Kupaʻa Hee. While I had already seen plenty of the critically endangered snails of the genus *Achatinella* in the lab, living out their lives in small plastic containers, this was to be my first encounter with a tree snail in an actual tree.

But as we entered the exclosure there was not a snail to be seen. Instead, at first glance the place looked just like any other patch of relatively healthy Hawaiian forest. Dave assured me though that there were plenty of snails to be found, it just took a little patience and some adjustment to one's mode of seeing—moving slowly, looking carefully, seeking out the particular trees and the other kinds of places that snails frequent. With a fair bit of Dave's assistance, I took up the search and was able to see a remarkable diversity of snails. As I had hoped, there were many of the larger, showy, tree snails with their patterned white and brown shells. But there was also a variety of less conspicuous ground-dwelling snails, making their lives within the decaying leaves at our feet.

I'll say a little more about the particular snail species in the exclosure in a moment. I'll talk about them one at a time, in the hopes that doing so might make it easier to keep some of the particulars of their tiny forms and lives straight in our heads. Each is, as we will see, fascinating in its own ways. My aim in returning to these snails below is to go beyond a focus on their varied shells—the most immediately apparent aspect of their diversity—to attend as best I can to the actual lives of these creatures.

There is a tendency in many snail discussions to focus almost exclusively on their shells, what Peter Williams has called "the safe, non-aversive, bit."[1] While a snail's shell, which begins to form at the outset of gestation, is very much a part of it—so much so that we should resist thinking of the snail as residing 'within' the shell and instead understand it as a being that takes form, in significant part, 'as' a shell—I aim to show here that these sturdy calcium carbonate structures are very far from being the most interesting thing about them.

On the day of my visit to the exclosure I was drawn into the lived worlds of snails. As I learned to search for them in the environment, and chatted with Dave about their habits and habitats, I began to realize just how little I knew about how snails perceive, navigate, and make sense of their surrounds, how they actually live their lives. I decided that day, spurred by my encounters in the exclosure, to rectify this deficiency in my understanding, and set about reading and interviewing leading snail biologists around the world. I discovered that slime is central to snail ways of life in both expected and unexpected ways. Their slime, as we will see, allows snails not only to move around but to construct a world of meaning by picking up on the chemical cues of their own and others' sticky secretions.

Traveling along these silvery pathways, this chapter is a foray into the worlds of snails—or at least an effort to say what we can about their general contours. In doing so, as the philosopher Brett Buchanan has outlined, my goal is not so much about getting inside the mind of the snail, "not about getting the animal 'right,' but rather about following the traces and paths of animals to see what they have to tell us, both about themselves and about our own selves."[2] As Hawai'i's snails face growing challenges to their continued survival, my hope is that attending to slime may provide some fragile possibilities for ongoing life, opening up new avenues for both conservation and appreciation.

There is something oddly poetic about this situation. Slime is often derided, seen as icky and gross; in fact, an aversion to slime seems to be at least part of why it is so difficult to convince many people to care about snails. So, it is fitting that this very same substance might be a pathway into something else. In this respect, it is worth remembering that slime is also generative: it is that "darkly vitalistic substance of life."[3] In the creation stories of both contemporary biology and Kanaka Maoli culture, slime is there at the beginning. It is the primordial ooze from which all life springs. Perhaps it is time for a new appreciation of slime and the diverse living worlds that it nurtures into existence. As the *Kumulipo*, an important cosmogonic Hawaiian chant, tells us:

The slime, this was the source of the earth
The source of the darkness that made darkness
The source of the night that made night.[4]

A BUBBLE OF PROTECTION

The first snail I encountered in the exclosure that morning, one that I managed to find all on my own, was a tiny creature. Only about 4 millimeters in length, it had a thin, translucent, shell that meant that if you looked closely enough you could just make out some of the flesh and internal organs beneath. Its pale body too seemed to lack the kind of firm, even if leaky, borders of more familiar snails: it was difficult to tell where flesh ended, and slime began. I asked Dave about it. "It's from the genus *Succinea*," he told me. And then, somewhat appropriately given my nascent interest in slime, he introduced me to the common nickname for these snails: "affectionately, we call it snot in a hat."

This snail, like all the others I encountered that day, inhabits a little bubble of protection in the middle of the forest that is made possible by the remarkable work that the exclosure fence does in keeping predators out. Rats and chameleons are excluded relatively easily: for the most part the curved lip at the top of the fence does the job; getting past it would require them to hold onto the slippery surface while moving upside down. A large, cleared area, a moat of sorts, around the exclosure prevents these animals from using nearby tree branches as bridges, while fencing beneath the ground stops them from tunneling under. Ongoing monitoring and baiting within the exclosure aims to kill any rats or chameleons that do manage to make the difficult crossing.

Keeping the carnivorous rosy wolfsnails out of the exclosure, however, is a much more complex proposition. But years of experimentation, alongside heartbreaking trial and error, have produced an intricate series of barriers that, for the most part, seems to do the trick. As a wolfsnail climbs up the outside of the fence, it first encounters the angle barrier, a metal flange jutting out downward from the main surface with an acute internal angle. As it proceeds forward, its shell will eventually hit this metal, stopping the snail in its track. This relatively basic piece of technology was developed and refined by escargot farmers in an effort to keep snails in, rather than out, of particular areas. It takes advantage of the fact that snails can't move backward, one of the many peculiar facets of their particular, slime-powered approach to locomotion (discussed further below).

If, however, our snail does somehow manage to scramble onto and over this obstacle, it will encounter the cut-mesh barrier, a small shelf about 10 centimeters wide that juts out at a right angle from the main fence. The underside of this shelf is lined with a spiky mesh. The wolfsnail must crawl across this surface upside down, where the difficulty is in maintaining enough adhesion with so little surface area to grip. Most snails fall off.

But if the snail does manage to make it over this second obstacle, it encounters a third: the electric barrier. Here it must cross several wires that are constantly pulsing a low-voltage charge, powered by solar cells. The charge is not lethal, but it delivers enough of a shock for the snail to contract its body, and thus lose grip on the fence and drop to the ground.

The barriers in place at the Palikea exclosure offer a formidable set of obstacles to a would-be snail predator. These barriers are the product of more than 20 years of experimentation and refinement. The first exclosure in Hawai'i was built by the state government in 1998, and was situated roughly 15 kilometers north of Palikea, within the Pahole Natural Area Reserve. It was Mike Hadfield who brought the exclosure idea back to Hawai'i, after a trip to the island of Mo'orea in French Polynesia. Mike had visited the island in the mid-1990s and taken an interest in the threatened Partula snails. Paul Pearce-Kelly and other staff from the Zoological Society of London had recently established a small exclosure to protect these snails from the rosy wolfsnail, which had been introduced there in the 1970s. The Moorean exclosure had a low fence; Mike recalled that "you could step over it." But while it was actively maintained it seemed to do the job.

This first Hawaiian snail exclosure was driven by the plight of the handful of remaning species of the genus *Achatinella*. These snails had been listed under the Endangered Species Act in 1981, following a petition for listing filed by Alan D. Hart that drew on his independent, island-wide, survey conducted over a five-and-a-half-year period. Of the 41 recognized species of this genus, Hart had been able to locate only 19, all of which he determined were either rare or extremely rare (only nine are thought to survive today).[5] Several other individuals and organizations made submissions to the listing process. Among them was Mike, who reported the results from his first in-depth population study of *Achatinella mustelina*, which commenced in 1974, including the rapid destruction of the study population with the arrival of wolfsnails in the area.

In the late 1990s, almost 20 years had passed since the listing of the *Achatinella* snails, and their situation had only deteriorated. While the captive lab that Mike and colleagues had established at the University of Hawai'i at Mānoa had become a home to small populations of some of these species, they were still steadily declining in the island's forests. Mike convinced the state government that with a few modifications the exclosure approach might help these snails to hold on a bit longer.

The first Hawaiian exclosure was a basic affair. The fence was constructed from corrugated metal sheets, and had a little roof that ran along its top edge, under which was a trough filled with coarse sea salt that the predatory snails wouldn't cross. Just to be sure, a two-wire electric barrier was also fitted, powered by a solar panel. While eventually the highly saline water being created by rain getting into the trough corroded the metal fence and created holes in the barrier, it worked well enough for long enough to establish the idea that this kind of in situ conservation might be a realistic option for Hawaiʻi's snails. After its completion, Mike told me, "we watched the snails vanish from all the trees around there, but our population inside remained."

In the years since, a total of 12 exclosures have been constructed across the Hawaiian Islands, with four more currently in various stages of planning and construction. Some of these facilities have been funded and built by the state government, others by the US Fish and Wildlife Service, but most have been constructed by the US Army in an effort to meet the military's legal responsibilities to stabilize endangered species on their lands (discussed further in chapter 5). The Army's work in this area is predominantly undertaken by the Oʻahu Army Natural Resources Program (OANRP). Today, the various exclosures dotted around Oʻahu are managed and run by either SEPP or OANRP, or jointly.

Irrespective of who is technically in charge of a given site, these two organizations collaborate closely, often sharing resources and expertise, as well as snails: populations found by one team that would be better housed in an exclosure run by the other, or in the SEPP lab, are often transferred. While it is the Army that has had the resources to construct most of the exclosures up to this point, the balance is shifting now as SEPP works to construct more of these protected areas on Oʻahu and on the island of Maui, which has the next highest number of recognized snail species after Oʻahu, including many that are thought to be similarly threatened but have been largely overlooked until recently.[6]

On the day of our visit to the Palikea exclosure, Dave had the additional job of trimming a patch of about 3 square meters of invasive grass, by hand, to look for signs of wolfsnails. Despite all the efforts made to secure the exclosure from these predators, it is still possible for them to make their way in. So, the SEPP team need to be constantly vigilant in their monitoring. Having offered to be helpful on the trip, I joined in with the work. For about an hour we crawled around on our hands and knees, cutting grass with small hand scythes. Each handful of cuttings, along with the grassy stumps left behind in the ground, had to be inspected for live wolfsnails or their shells, as well as for subtler traces like their slime trails or eggs (which Dave described to me as looking quite a lot like Tic Tacs, bright white and elongated). Fortunately, on that day we encountered no disturbing signs. As far as anyone knows, the Palikea exclosure has been predator-free for several years now.

During the initial establishment of Palikea, days and days of painstaking inspection work like this was done, combing through grasses, bushes, and trees, to remove wolfsnails and their eggs. In comparison to Hawai'i's remaining snail species, wolfsnails are distinctly large, their pinkish-colored shells usually measuring somewhere between 5 and 7.5 centimeters in length. Even still, they can be pretty difficult to spot among the vegetation and so there can be no guarantee of finding them all through this kind of visual inspection. In fact, Dave suspects that the occasional wolfsnails found inside over the years might be descendants of snails that have been lying low all along, rather than incursions over the fence. As a result, some of the more recently constructed exclosures have adopted a "scorched earth" approach in which all vegetation but the odd tree is removed, and the area completely replanted. Of course, this drastic tactic requires years of recovery time before the area is suitable to release snails into. There are no simple solutions here. Keeping Hawaiian snails alive in the forest, even behind a carefully designed fence, is incredibly hard work.

Monitoring for wolfsnails inside exclosures is particularly important because one individual can quickly become many. Like most land snail species around the world, wolfsnails are hermaphrodites. A single individual entering the exclosure might be able to lay hundreds of eggs within a single year. Each of these eggs might grow up into a snail that is able to start reproducing within its first year, laying hundreds more eggs of its own.

This reproductive capacity is a key part of the conservation challenge. In contrast to the prodigious productivity of their predators, many of Hawai'i's snail species, especially the larger tree snails like those of the genus *Achatinella*, are long-lived and slow reproducers. In fact, this is probably something of an understatement. To the shock of Mike and his colleagues conducting the first sustained studies in this area, these snails frequently live for 15 years or more and don't reach sexual maturity until they're around 5 years old.

When they do start reproducing, instead of egg clusters they give birth to a few live-born young per year.[7] As Mike summed it up for me: "their reproductive life histories are much more like those of a bird or a mammal than they are like most other invertebrates."

It seems likely that these larger tree snails have been significantly shaped by life in an environment with little or no predation. In fact, one theory suggests that they may even have taken to the trees—very slowly, through successive generations of adaptation—to avoid their only significant predators, large ground-dwelling birds that are now all long extinct. In the absence of predators, their reproduction was able to slow right down. As Mike put it, discussing the initial discovery of the specifics of *Achatinella* reproduction: "No other such snail life history had ever been reported, and we found it repeated for population after population and species after species of Hawaiian tree snail."[8]

The drama playing out in this exclosure is one that falls pretty easily into familiar story lines about villainous invasive species, destroying vulnerable native species and their ecosystems. There is some truth to these characterizations, which is a big part of what makes them so compelling. The wolfsnail is the most significant factor in the decline of a range of native snail species in Hawai'i today. But, of course, the wolfsnail is just doing what predatory snails do, after having been deliberately brought to these islands by careless people.

The rosy wolfsnail (*Euglandina rosea*) was introduced to Hawai'i from Florida in 1955 as part of a poorly planned initiative by the territorial Department of Agriculture. The idea was that these predatory snails might help to control the giant African snails (*Lissachatina fulica*) that had been earlier introduced and were now deemed by some to be significant horticultural and agricultural pests. At around the same time, a similar story unfolded across the Pacific and around the world, with wolfsnails being introduced to a range of other islands. In Hawai'i, as in many of these places, not only did these introduced snails fail to reduce the populations of their intended targets; they ended up decimating the locals. The wolfsnail has now been implicated as a major contributor to numerous snail extinctions on islands in the Pacific, Indian, and Atlantic Oceans.[9]

It is clear that very little research was conducted to evaluate the potential impacts of wolfsnails in Hawai'i. In fact, a variety of other species was also released in this period. A total of 19 snail and 11 insect species were brought to Hawai'i within just 15 years as part of the biocontrol program for the giant African snail, with most of them subsequently being released into the environment.[10] During this period, very few trials were conducted to understand whether they would be at all effective in reducing the overall abundance of their intended target. Even more concerningly, there is no evidence that any meaningful investigations were undertaken whatsoever on the potential "nontarget" impacts of all of these new snail predators being brought to Hawai'i.[11] For the people running these programs, the incredible endemic snail diversity of these islands didn't even register as something to be given a second thought.

In fact, it now seems that the snails we've been calling rosy wolfsnails (*Euglandina rosea*) all these years may actually be members of two distinct species, perhaps even three or more. While molecular evidence unavailable in the 1950s has enabled part of this new understanding, the scientists who made this discovery noted that there are also clear morphological (physical) differences between the (possible) species that might have been observed through more careful attention.[12]

Clearly "the snail formerly known as *Euglandina rosea*" has had a very significant impact on gastropod diversity in Hawai'i and around the world. Without wanting to diminish this fact, it is also important to note that there is a tendency in some conservation discussions to scapegoat introduced species in a way that locates the problem too narrowly, focusing in on the voracious appetites of problematic pests. These kinds of understandings often conveniently gloss over the role of some humans in these arrivals (usually driven by a combination of carelessness and nationalistic, financial, or political agendas), while also ignoring the ongoing processes of habitat loss and disturbance which frequently amplify the impacts of newly arrived species.[13]

As we will see below, and in the chapters that follow, the decline of Hawai'i's snails is a much longer, larger, and more complex story than any simple, singular villain. As the biologist Claire Régnier and her colleagues put it in their discussion of recent snail extinctions in Hawai'i, the rosy wolfsnail "was the precipitous cause of these extinctions but populations of these endemic snails were already weakened by decades of habitat destruction, overcollecting, and predation by other accidentally introduced species."[14] In other words, the wolfsnail is, as Mike has noted, "the final blow" in a long history of other impacts.

WORLDS OF SLIME

The first of the larger tree snails that I encountered in the exclosure was a group of about seven *Achatinella mustelina*, resting in the midday heat. The mostly white shells of this group stood out among the greens and browns of the forest, but it wasn't their shells that initially drew my eye. These particular snails stood out because they were clumped together on a piece of fluorescent pink plastic tape. The tape itself was tied to a large tree at about eye height. It had been placed on the tree to mark it as one in which snails were present; the snails had in turn decided that the tape itself was a good place to estivate, the process whereby a snail pulls into its shell and seals up the shell's aperture (opening). This seal is created by a thin layer of the snail's own slime (called an epiphragm), which dries around the edges of the aperture into a hard glue. This glue allows the snail to firmly attach itself to an external surface—in this case the pink tape—so that the snail might retain its moisture.[15]

As I stood and watched this inactive cluster of snails, I found myself more and more curious about their largely unseen routines. Did they usually gather together on this pink tape? When they moved around at night, how far did they go? If they made their way back to this spot each morning, how and why did they do so? What kinds of understandings, social relationships, and evolutionary imperatives grounded this pull to aggregate in a clump, to spend the day resting alongside other snails?

The first thing to note in exploring these questions is that snails rely on very different modes of sense perception from those that dominate (most) human lives. To begin with, snails are pretty much blind. This situation comes as a surprise to many people. After all, most land snails have quite conspicuous eyes, or at least eye-like bulges, readily perceptible on the uppermost tentacles that extend from their heads (they're called "tentacles," not "antennae," in gastropods). But snail eyes are thought to be primarily involved in light sensing, helping to determine the time of day and so regulate circadian rhythms. At best, snails can probably discern blobs of dark and light—a situation that makes some sense given that they lead primarily nocturnal lives. Alongside their all but absent vision, snails are also thought to be largely deaf.

In place of these senses, snails' worlds take shape primarily through chemoreception. The closest approximation within the human sensory repertoire is our own senses of smell and taste, both of which pick up chemical signals in the environment. But for a snail, these chemosensory faculties are much more finely tuned. It is here that those same tentacles on a snail's head, packed with sensory neurons, really come into their own. For most terrestrial snails, there is a division of labor here with the two "superior" tentacles—positioned higher on the head and larger in size—focused on long-range sensing, and the two "inferior" tentacles being primarily used to smell/taste things directly in front of the snail.[16]

The world is saturated with chemical cues for those with the capacity to taste them. Snails pick up on a range of chemicals, but their own and others' slimy secretions are a particularly important source of information. Dave explained to me that it is most likely by attending to slime that the snails on the pink tape found their way back to that spot each day, a process referred to by biologists as "homing." They might do so by reading the slime trail laid down by their own or others' past movements—indeed, smell-tasting these trails is one of the principal tasks of their inferior tentacles. But they may also make use of their superior tentacles in this task, sensing from a distance and homing in on the cumulative slime that has built up in this place.[17] Dave speculated that the plastic tape might even retain these chemical cues for longer than tree bark and other permeable surfaces, thus offering a stronger and more reliable home site (as well as a nice tight seal for estivating snails).

This was the slimy path by which I made my way into the fascinating lives of Hawai'i's snails, their modes of perceiving, navigating, and inhabiting their worlds. When we turn to the contemplation of slime, however, the first question must surely be: Why slime at all? Plenty of other organisms seem to do just fine without producing and secreting large quantities of the stuff. Scientists tell a range of stories about the origins and functions of snail slime. Technically called "pedal mucus," it is composed mostly of water with traces of carbohydrate–protein complexes that are responsible for its characteristic stickiness.[18] As the term "pedal" implies—an adjective referring to something of or related to the foot—this slime is thought to have originally been an adaptation for locomotion. It works, however, by providing not a lubricant, but rather an adhesive. There is definitely something counterintuitive about this situation. As the biologist Mark Denny put it in a now classic paper: "How can an animal with only one foot walk on glue?" The answer, it seems, rests in both "the unusual mechanical properties of the pedal mucus" and the wave-like motion of a snail's foot as it inches along. The sticky mucus allows the snail to use its foot like a "material ratchet, facilitating forward movement, but resisting backward movement."[19]

Importantly, this slime also opened up a three-dimensional world for gastropods. It is by virtue of its adhesive properties that snails are able to travel vertically and upside down, and in the case of marine snails also to inhabit wave-swept environments.[20] Without slime, snails with their bulky shells would occupy a much more constrained environment. At the same time, slime expands the range of climatic environments a snail can occupy through the role it plays in allowing them to stay moist and seal up in warmer and drier weather.

It is these possibilities that are thought, at least originally, to have led to the evolution of this peculiar mode of locomotion. These advantages justified the very high metabolic cost of producing and secreting large quantities of nutrient-rich slime: so high, in fact, that Denny has calculated that traveling via slime is an order of magnitude more expensive, in terms of the energy invested, than any other mode of locomotion.[21] Fascinatingly, it seems that snails have developed a means of making the most of this slimy investment in travel: they can and do *reuse* their own and others' trails to significantly reduce the need to secrete mucus.[22]

But this is not where the story ends. Over the roughly 550 million years that gastropods have inched their way around this planet, this locomotive slime has taken on a range of additional purposes, and indeed meanings, in the lives of snails. All over the world, in aquatic and terrestrial environments, snails are simultaneously laying down and reading trails of slime. As a result, this sticky substance has come to form a key part of how they navigate and make sense of their environment, spatially and socially.

Snails, like all living beings—from bacteria and plants to humans—inhabit their own particular worlds of meaning. Each is, in their own ways, sensitive to their environment, taking in specific information and responding. It was this insight that lay at the heart of the important work of the early twentieth century Baltic biologist Jakob von Uexküll, focused on what he called an *Umwelt* (or "surround-world").[23] In contrast to biological work that emphasizes the purely biophysical environment of an organism—its habitat or niche—in using this term Uexküll aimed to draw our attention toward the very different worlds of meaning that living beings occupy as a result of their unique ways of sensing and making sense. A snail, tuned in to an array of chemical signs that we simply cannot comprehend, inhabits a different world—a particular *Umwelt*—in which different entities and possibilities emerge and move to the fore.

In inviting us to imagine these diverse worlds, Uexküll insisted that meaning cannot be reduced to language, as is so often assumed, and as such that it is not just language-using species like humans that make and exchange meanings. From his perspective, the lived world is one that is, in an important sense, *crafted* by the organism through their particular embodiment: their modes of perception and their specific needs, cognition, desires, and life histories. While it is not possible to enter into another's *Umwelt*—to see through their eyes, or in the snail's case to smell through their tentacles—glimpses (or wafts?) of insight can be pieced together to enable us to understand just a little bit more of their lives. Uexküll described these imaginative excursions as "forays" (*Streifzuge* in his native German).

In the case of snails, slime trails have often offered important entry points for these kinds of efforts. There have now been decades of scientific experiments on "trail following" among a wide range of gastropod species. Attending to snails as they attend to slime, scientists have been able to develop a better understanding of the worlds of meaning that snails construct, inhabit, and share with one another.[24]

In 2017, in a small laboratory on the campus of the University of Hawai'i at Mānoa, three researchers conducted an experiment that aimed to ask several local species of snails about the slime trails that interest them. Over a period of about 6 weeks, Brenden Holland and two coauthors—Joanne Yew and Marianne Gousy-Leblanc—carefully observed the slow movements of hundreds of snails. Each one was placed on a Y-shaped branch and given the opportunity to either follow, or not, a previously laid-down slime trail. The experiment worked like this: in each case a fresh branch of the 'ōhi'a plant was used, a favorite host tree for the snails. A "marker" snail was placed near the diverging point of the Y and allowed to make its way along one or other of the forked branches. This snail was then removed, and another "tracker" snail was placed back near the starting point.

Repeating the experiment over and over again, with hundreds of snails of various ages from six endemic species, a clear pattern emerged. Whether or not a snail follows another's trail depends on the relationship between the two. If they are of different species, then there seems to be no deliberate following: tracker snails chose one of the two paths open to them seemingly at random. The same was true for snails of the same species (conspecifics) when a keiki (juvenile) was involved, either as a marker or a tracker. But when two adult snails of the same species were observed, again and again, the tracker opted to follow the marker. This didn't happen every time, but during roughly 78% of experiments with conspecific adults, following took place. The researchers called this "a highly statistically significant" result. In addition, frequently the tracker snail didn't just choose the same fork in the branch; it followed precisely in the footstep of the marker, repeating the circuitous spiraling route that this snail took as it went both along and *around* the branch.

To rule out the possibility that these tracker snails might be relying on visual and long-range chemical signals as part of their decision making, the researchers also introduced a control study in which they placed a resting snail at the end of one of the two forked branches. They concluded that this did not influence snail decisions. The trackers, it seems, were relying exclusively on the slime trail itself to determine whether or not it was worth following. This finding was further reinforced by mass spectrometry analysis of the slime, which revealed distinctive differences between adult and keiki mucus. Specific pheromones were found only in the slime of the adults, and the researchers concluded that they were likely a key part of its communicative power, enabling or even enticing following.

It is impossible to say with certainty what motivates this following behavior. It seems unlikely that it is purely about the kind of locomotive efficiency mentioned earlier. If it was, it shouldn't matter what the species or the age of the snail being followed was. It is certainly possible that snails are following one another for a range of reasons. Perhaps in the absence of the appropriate pheromones the trails of juveniles and other species are unintelligible; perhaps adult conspecifics are preferentially followed because they're a more reliable guide to food. Perhaps it depends on the time of day and the specific needs and inclinations of the tracker. Like the philosopher Vinciane Despret, I think that holding open room for diverse possibilities—rather than rushing to provide *the* explanation for a behavior—is a vital part of engaging "politely" with other species, and ultimately of doing good science.[25] Nonetheless, whatever else is going on here, it does seem very likely that finding a reproductive mate is at least part of the story. Such an explanation would align with the fact that snails preferentially follow adults of the same species. As these snails are all hermaphrodites, any individual of the same species is a potential reproductive partner (importantly, while some Hawaiian snail species may be capable of self-fertilization, this practice seems to be rare, or, at least, it has rarely been observed).

Similar studies have been conducted on snail and slug species all over the planet to learn more about their slimy inclinations. This research has documented a range of other fascinating dimensions to trail following, while also reinforcing the hypothesis that it plays a role in finding a mate. In one study, pond snails (*Lymnaea stagnalis*) preferentially followed potential *new* partners—that is, snails that they hadn't recently copulated with.[26] In other studies it was found that snails are able to acquire important information about potential mates through their slime trails and are reluctant to follow individuals who are poorly fed or have been rendered sterile by parasites.[27] It seems that there is a wealth of information about others in their mucus for those who know how to taste it.

Alongside these social interactions, we have seen that slime also provides a means of spatial orientation for snails. The estivating bodies of all those *Achatinella mustelina* clumped together on the pink tape remind us that snails are "homebodies" in some crucial ways. In contrast to the common view that a snail "carries its home on its back"—like an itinerant wanderer, presumably able to set up camp anywhere—most snails are actually pretty tied to particular home places. In Hawai'i, these are usually the spots where they rest during the daylight hours between their long nocturnal feeding excursions. As previously mentioned, slime is central to the story of this homing capacity, too. Snails might follow their own or others' slime trails directly back to their home, or they might read long-range chemical cues, including the accumulated mucus at a home site.

The specific homing abilities of Hawai'i's snails have not been experimentally tested. But if they're anything like those of the other species that have, then they're up to some remarkable things. A now classic study carried out by Carl Edelstam and Carina Palmer put common European land snails (*Helix pomatia*) through all sorts of obstacles to find their way home.[28] The researchers asked: Will they head home if they are moved a few meters away? What about 70 meters? Will they cross hot, dry gravel and other unsuitable habitat types to get there? What if they're collected, kept in a bag in an apartment in a nearby town for a couple of days, then released about 30 meters from their homes? In all of these cases, once released the majority of the snails quickly headed off in the right direction.

The few other studies that have been conducted on snail homing in the years since—focused on various aquatic and terrestrial gastropods—have shown that this behavior is widespread. There is, as such, every reason to believe that the same capabilities are informing the homing behavior of Hawai'i's snails.

But why head home in this way? On one level the answer is probably quite simple. As Edelstam and Palmer summed it up: "on the average, an organism with little ecological plasticity will always have a greater chance of survival in the area where it has already persisted for some time with success."[29] A reliable home place provides refuge from predators and from the weather, while also being close to suitable food and potential mates. If you've already found a place like this, stick with it. Some of these benefits might just come from its being a good spot, but in other cases the benefits actually arise from the aggregation of many snails: for example, hanging out in a group reduces one's chances of being eaten and can also help to retain moisture.[30]

It is also worth noting that snails seem to *like* gathering together.[31] Admittedly, the social likes and dislikes of snails are a particularly difficult topic to gain insight into. But experiments conducted over the past decade have shed some light on this topic, in particular, work in the laboratories of Sarah Dalesman at Aberystwyth University in Wales and Ken Lukowiak at the University of Calgary in Canada. The principal focus of their experiments is learning and memory in pond snails.[32] Like other similar studies, the research being done in these labs has found that snails are capable of learning about new foods or potential predators, and adapting their behavior accordingly. But Sarah and Ken have also worked to understand the social factors that influence the ability to pick up and retain new information. Fascinatingly, they have found that snails that are either isolated or overcrowded experience stress.

While we can't really know anything about what this social stress *feels* like to a snail, we do know that it produces negative metabolic and cognitive changes (like impacts on shell development and memory formation). In the case of overcrowding, we also know that these impacts are not simply the result of being physically constrained. In an experiment that Sarah and Ken conducted together, snails placed in a confined space with a lot of empty shells that similarly restricted their movement, or with snails of a different strain (an unrelated group from the same species), were not impacted in the same way. The physical constraint had to come from being crowded by actual, living snails of one's own social cohort. Fascinatingly, as Ken explained to me, slime seems to be a key communicative medium here, too, providing information to the snail about who is around them, or who is absent. In other research Sarah and Ken have found that isolation induces social stress in a similar way to overcrowding.[33]

There remains a great deal that is not well understood about all these social interactions, but it is clear that snails are socially aware beings in their own snaily ways. Alongside all of the other things they're paying attention to and learning about, they are tuned into a (slime-mediated) social context and seeking out some arrangements and not others. It is likely that Hawai'i's snails are negotiating such social worlds. If this is the case, then those snails gathered on that pink tape are drawn together not only by the capacity to smell, taste, and follow slime but also by some sort of inner imperative to be with others of their kind.

But these are not the only slimy stories unfolding in Hawai'i's forests. Importantly, it isn't only the endangered, native snails that are reading and following the cues laid down in mucus. In fact, it is precisely this capacity that makes the rosy wolfsnail such an efficient and devastating predator. Alongside following the trails of conspecifics to mate, these carnivorous snails are also able to use them to identify and track their next meal. A small study conducted by Brenden and colleagues in 2012 showed definitively that wolfsnails are using slime trails to track Hawaiian tree snails, in this case *Achatinella lila*. Of course, in this study the wolfsnails weren't allowed to actually catch and consume the endangered snails, but they clearly picked up on their trails and followed them.

Fascinatingly, this study also confirmed the long-suspected view that wolfsnails *preferentially* target the Hawaiian snails over the giant African snails they were introduced to eat. In 15 out of 20 trials, using a similar Y-branch test, wolfsnails chose to follow Hawaiian tree snails over giant African snails.[34] Unlike most snails that primarily use their lower, inferior tentacles for trail following, carnivorous wolfsnails also possess specialized elongated lip tentacles, that look quite a bit like a handlebar moustache. These lips are thought to be the primary means by which they taste slime trails.[35]

While wolfsnails are able to consume prey that are larger than themselves, all of the remaining native species in Hawai'i are smaller than an adult wolfsnail. When their prey has been located, wolfsnails make use of another specialized appendage, an eversible proboscis inside the mouth. The wolfsnail holds its prey with its foot, placing its own mouth at the opening of the other snail's shell. Mike Hadfield explained what follows to me: the wolfsnail "forcefully everts its proboscis into the prey's shell. At the tip of the proboscis are powerful jaws, the radulas, and the openings of powerful digestive glands. It literally tears and digests its prey within its shell within minutes and swallows it while it retracts the proboscis." Smaller snails are often simply eaten whole, shell and all. In fact, smaller snails seem to be a preferred source of food, with their shells perhaps providing an important source of calcium.[36]

Importantly, wolfsnails are able to adapt their trail-following behavior to take advantage of new prey. This situation is clear in Hawai'i where the native snails were quickly added to the menu by the newly arrived carnivores. But a study at Delaware State University found that they might be even more adaptable than this. It seems that wolfsnails can quickly learn, after only one or two trials, to associate entirely novel chemical cues with prey and to preferentially follow those cues.[37] Wolfsnails are, in other words, highly efficient and adaptable predators.

Slime, that humble mucus, turns out to be a core component of snail worlds. Paying attention to its stories draws us into a fuller sense of the remarkable lives of these creatures. In this way, somewhat counterintuitively, attending to their slime helps us to see that snails are themselves, as Sarah Dalesman succinctly put it to me, more than just "bags of slime." They are beings that inhabit remarkably rich *Umwelten*, worlds that are quite literally slimed into being. This is true in the material way in which slime opens up a space of three-dimensional movement, allowing snails up into the trees and out into diverse environments. But it is also true in that subtler sense in which slime is a complex communicative matrix, forming a key element of a snail's landscape of meaning: mapping out a terrain of familiar places, of potential mates, of homely aggregation.

This situation reminds us that the worlds of perception that animals inhabit are not pre-given: they are as much *crafted* as they are *encountered*. As they move through the landscape, snails layer meaning into their own and others' worlds. As the poet Gary Snyder has put it: "Other orders of being have their own literatures."[38] In the most vital of ways, slime is a storied, world-forming substance. But of course, these snail worlds are today also being destroyed. At least in part, they are being slimed *out* of existence, too. With the aid of their fine-tuned lip tentacles, wolfsnails are able to reread and repurpose the very substance that wove together the complex worlds of Hawai'i's snails to efficiently unravel them from the inside.

WANDERING

One of the most striking snail species I encountered in the Palikea exclosure was *Amastra spirizona*. This species, with a beautiful brown, conical shell about 1.5 centimeters in length, is now thought to be found nowhere else. Once widely distributed across the Wai'anae Range, in 2015 Dave and his team noticed that they were disappearing. "They got bottlenecked down to about 30 snails," he told me. "It got to the point where we just had to take them all out."

Collected from the forest, the snails were carefully packed into plastic containers which were then transported by foot to a nearby summit where they could be air-lifted out by helicopter. Eventually, they found their way here, to the Palikea exclosure. But they weren't just released into the exclosure at large. Instead, a square wooden box about a meter in size with wire mesh on one side was built for them inside the exclosure. Unlike the *Achatinella* tree snails that live up in the branches, *Amastra spirizona* is one of the island's detritivores, a consumer of dead and decaying leaf matter on the forest floor. And so, their makeshift cage has been placed on the ground, complete with decaying branches and leaves that can be regularly topped up for them.

On first encountering these snails I assumed that the purpose of their box was to provide them with an extra layer of protection, in case a wolfsnail or a rat managed to make its way over the exclosure fence. But I soon discovered that its primary function is much less dramatic. As Dave explained to me, it is intended simply to keep the snails together: "A lot of times when you translocate snails, they wander pretty widely. . . . So, we were worried that with so few they wouldn't find each other, they'd all just wander away. This was a way to contain them and keep the breeding adults together." At the same time, the wire mesh on the side of the box allows any keiki produced from that reproduction to disperse into the exclosure beyond. Happily, this is exactly what has happened: at the time of my visit the population of *Amastra spirizona* was thought to be around 150 snails.

Knowing what we do about the way in which the social and spatial worlds of snails are crafted in slime, it should perhaps come as little surprise that when they are picked up in one place and translocated to another, sometimes miles away, they don't quite seem to know what to make of things. And so, they head off in all directions. As Mike neatly summed it up: "if you move them, they don't stay put; you make wanderers out of them." While part of the challenge that this box responds to is simply one of a small population of animals that can't always readily find one another when spread out, it is also a technology animated by something peculiar to the lifeworlds of snails, this tendency to wander in a new environment.

But *Amastra spirizona* is not the only species for which this behavior is a conservation concern. Dave explained that many of the tree snails that are newly released into an exclosure exhibit this kind of wandering, too. "We'll find the nicest tree and we'll put them in there . . . but we find through our monitoring efforts that these snails will disperse widely across grass, and they'll end up in plants that we don't normally see snails in." This behavior is a worry both because the snails reduce their chances of finding one another, and because they often move into what seem to be suboptimal plants—at least to the extent that well-qualified humans are able to assess these things.

Some of these wanderers are snails that have been moved directly from forests elsewhere on the island to be released into an exclosure; others have been born and raised for their first years of life in the laboratory. Either way, they are often being moved miles away into an unfamiliar terrain. In this context, they probably have no sense of where "home" is, so cannot set off in that direction. Fascinatingly, it seems that even in cases like this, snails prefer not to stay put. In their experiments in Sweden with the snail *Helix pomatia*, Edelstam and Palmer found that, when given the opportunity, displaced snails headed off in "a fairly straight way." While most of the snails in their study were indeed heading toward home, even those that weren't took a direct line of travel, leaving this unfamiliar place behind for good. This movement, the researchers noted, contrasts significantly with undisturbed snails, who tend to "rove about [seemingly] at random," venturing out from and back to the same home place each day.

It is unclear what motivates these straight-line movements. Snails certainly seem to be heading either toward or away from something in particular, but we can't know that this is the case. Are these relocated snails lost and disoriented, fleeing a site of confusion? Are they confidently heading toward home, rightly or wrongly? Are they perhaps exploring, moving in a straight line until they encounter something that makes sense, perhaps the slime trail of another member of their species, to draw them into an existing world that they can begin to layer their own trails and meanings into? Or perhaps all or none of the above? Whatever their motivation, the key point is that relocated snails tend to quickly abandon the site they are dropped down in.

This situation is potentially even more complex for those snails that have been reared in the lab and then released into an exclosure. Slow-growing species like the *Achatinella* tree snails spend the first 5–7 years of their lives in small plastic containers stocked with branches and leaves of native plants. As discussed in chapter 6, these containers are sterilized and their vegetation changed every 2 weeks, both to provide fresh food and to prevent the potential establishment and spread of pathogens. Dave worries that in these conditions juvenile snails may have simply never learned, or even have unlearned, how to draw slimy meanings into and out of their landscapes.

There is very little research that might shed any light on this situation. One study conducted by Sarah and her colleague James Liddon, however, has found behavioral differences between snails reared in a lab and those collected from the environment. They showed that lab-raised snails that had been isolated for a time were less likely to try to follow other snail tracks. There is no clear explanation for this difference, but the finding does raise the possibility that captive rearing might alter snail behavior. As Sarah pointed out to me, while we have long known this to be the case for mammals and birds, there is as yet very little understanding about the impacts of captivity on invertebrate behavior. Perhaps in some cases the wandering witnessed in the exclosures is partly a product of snail upbringing, leading them to avoid following and aggregating with others.

The term "wander" implies purposeless movement. As the *Oxford English Dictionary* puts it: "To move hither and thither without fixed course or certain aim." As we have seen, it is just as likely that these snails are heading home, fleeing, or even exploring. Dave and Mike know this. In their description of these snails as wanderers it seems to me that they are gesturing toward the inscrutable, toward that portion of another's world—of their way of being and sense-making—that we cannot access or understand. But this situation does not mean that we can know nothing at all. As we have seen, there is much that can be said, better and worse hypotheses and interpretations, that can provide us with at least the faint scent of a meaning.

At the same time, these wanderers remind us of the vital importance of this ongoing effort to better understand the way in which snails attach meaning to their worlds if we are to have any hope of holding open room for them in the forest.

As the conservation situation of Hawai'i's snails deteriorates, their worlds of meaning are increasingly fractured. The philosophers Carlo Brentari and Matthew Chrulew have drawn our attention toward the possibility of extinction arising as a process of undermining the *Umwelt* of a living being. As Brentari has put it: "safeguarding biodiversity does not mean so much preserving 'the animals' as defending the *Umwelten*, the semiotic, perceptive and operative worlds in which life unfolds."[39]

Snails living beyond the exclosure, many of them in shrinking populations, perhaps experience this fracturing of their *Umwelten* as they come to inhabit less slimy worlds, with fewer pathways of meaning and sociality. As Brenden Holland has noted, in a "complex three-dimensional arboreal habitat" reductions in population make life, and reproduction, that much more difficult. It seems likely that slime networks really only "work" if snails occur at minimum densities. Beyond this, they simply break down. This is, in a sense, the problem faced by *Amastra spirizona*, but writ large throughout the forest, where there are no boxes to keep shared worlds together.

Similarly, for those snails living their lives within the confines of the exclosure and the lab, where immediate survival requires greater levels of intervention and disturbance on the part of conservationists, the possibilities for snails to craft the kinds of stable social spaces that enable aggregation and mating are perhaps being undermined. Understanding these dynamics might be vital to better conserving and living with snails. As Brenden has noted, reflecting on the conservation implications of his own research on trail following among local snails: "Detailed information regarding mate finding could prove useful to resource managers, especially in cases where density of arboreal populations can be manipulated."[40]

Close attention to animal learning and behavior is still not particularly common in many conservation projects, especially when it comes to the conservation of invertebrates.[41] But the more attention is paid to them and their experiential worlds, the more complex we are discovering their lives and needs to be. As Sarah noted: "Invertebrates in general are capable of a lot more than we potentially give them credit for. One of the problems we have, because they are so different from us, is learning to ask the questions in the right way. We have to put ourselves into the body of a snail or the body of a bee and try to think about how they perceive their world; what's going to be of importance to them; what they're going to respond to; and then how to ask them the right question. How to ask them, 'does this matter to you?'"

Dave and his team are determined to find ways to productively pose these questions to the snails, without causing further disruption to their lives through intensive experimentation. To this end, they are carefully trying out some different management approaches—altered regimes of release and confinement—and then monitoring to see what impact they have. Of course, the *Amastra spirizona* box is one such intervention. Developing this idea further, they hope to create similar spaces of confinement for tree snails and others released into the exclosure. These would be caged areas, "almost like a little aviary," in which, as Dave put it, they would be "forced to stay and develop a home range." During a period of weeks or months of containment, they would be able to layer their own slimy forms of meaning into their worlds. When the cage is removed, more of them might stay put. At least that is the hope.

Meanwhile, Dave is also planning to start releasing some lab snails when they're a bit younger. These young snails are smaller and more susceptible to drying out, but he hopes that if their wandering is being exacerbated by their lab-rearing, these behaviors might not have developed as strongly yet. These minor alterations to snail management, coupled with ongoing monitoring, might allow the team to better understand how and why these particular snails wander—and so to find ways to ensure that it doesn't undermine their chances for ongoing life.

As we prepared to leave the Palikea exclosure that day, I did the rounds to take a last look at some of the snails. Among them was a little group of *Laminella sanguinea*, a species that is not formally listed as endangered but widely considered to be by experts. *Laminella sanguinea* have a conical shell about 1.5 centimeters in length that is a vibrant red in color (as their Latin name implies) with a distinctive lightning-bolt pattern running from apex to aperture. Without a doubt they have one of the most visually stunning shells among the Hawaiian snails. But I only know about their beautiful shells because Dave told me about them, and I later looked up some photos. When I encountered *Laminella sanguinea* in the exclosure, like every other person who is lucky enough to see them out in the world, their shells were hidden beneath what looked like gray-brown dirt or mud. In reality, however, it was a layer of the snail's own excrement, secreted onto its shell. We don't know why these snails developed this particular practice; theories include camouflage from predators and thermal regulation, but neither is particularly compelling.

With its distinctive excremental proclivities, *Laminella sanguinea* reminded me that despite the many similarities between snail species for the average human observer—myself included—each of them is a unique form of life. Each of them is a mode of inhabiting and so of crafting a world. It is these *worlds* that are placed at risk by, and ultimately lost to, extinction. Beyond the concrete conservation measures that might be opened up by paying attention to slime, I am convinced that this sticky substance also creates new opportunities to appreciate gastropods. Reduced to shells or pests, the fascinating lives and worlds of snails all too often slip from view. But if we pay attention, snails are following, they're aggregating, they're homing, they're learning about and becoming stressed by changes in their social and physical environments. In short, they are up to some pretty remarkable things.

My point in noting this is not to imply that snails are "just like us" and so worthy of our consideration. They aren't. While we might share some cognitive and perceptual capacities, others are very different, incomprehensible even. Instead, the point of paying attention in this way is to learn, as best we can, to understand and appreciate snails on their own terms. To some extent, of course, we cannot escape our own anthropocentrism, just as we cannot escape our particular hominid visual-brain system; it shapes and constrains us and our experiences (even though there is some flexibility and diversity in its forms). But there is an important difference between the recognition of an unavoidable *epistemic* anthropocentrism and a much more problematic *ethical* one.[42] From within the confines of my necessarily human mode of knowing and inhabiting the world, my *Umwelt*, I can endeavor to make room for and appreciate others. Not because they are like me, but rather because they are each engaged in their own, unique, mode of life. In so doing, we might avoid what Eva Hayward has referred to as the simplistic and problematic "misalignment of empathy with the possibility of familiarity," opening up space for the valuing and appreciation of diverse forms of, sometimes radical, otherness.[43]

Through their distinctive modes of being, each of these snail species— indeed each individual member of each species—crafts its own *Umwelt*. To appreciate this situation is to be brought into an encounter with another layer of the tragedy of extinction. It is to come to appreciate, as Vinciane Despret has noted, that it is an entire world that is lost in extinction: "every sensation of every being of the world is a mode through which the world lives and feels itself . . . [and when] a being is no more, the world narrows all of a sudden, and a part of reality collapses."[44] From such a perspective, our contemporary period of incredible biodiversity loss must be recognized as one in which, as the philosopher Eileen Crist has noted, alongside destroying species, habitats, and populations, we are deleting countless experiential worlds that are "elaborated through emotion, intention, understanding, perception, experience—in other words, through varieties of aware beings shaping and adorning the world-as-home."[45]

While many Hawaiian snail species and their worlds have already been lost in this way, others still endure. These tiny beings are still out there in the forest, perhaps behind an exclosure fence, perhaps not. Either way, they are doing their best to layer meaning into and draw it out of these landscapes. But in many places this possibility must be becoming more and more tenuous as their numbers decline and their lives and landscapes become increasingly disturbed and fractured. The wandering snail is emblematic of this process of loss: a living being that has been severed from its world of meaning, made restless by the absence of those markers that designate a place as familiar, as home. But while the *Umwelten* of these beings must always remain partially unknowable to us, this does not—it *must* not—place them beyond the realms of our curiosity and our regard. They are worlds that there is still time to learn to make room for; still time to learn to appreciate in the silvery network of paths that weaves its way through the Hawaiian forest in those rare places where the snails still persist.

2 THE DRIFTERS
The Mysterious Deep Time Movements of Snails

We wove our way steadily upward. Brenden was driving, so I took the opportunity to look around. The road was steep and winding, in places snaking back and forth on itself just to make it up the hill. For the most part we were surrounded by forest on all sides. Occasionally, though, we passed through an area without trees, close to the cliff edge, and it was possible to see all the way back down to the sprawling suburbs of Honolulu, and the ocean beyond. As we neared the top of Tantalus Drive, the houses had mostly given way to forest with only the odd one or two dotted along the road. The hill we were making our way up was once called Puʻu ʻŌhiʻa, but is generally referred to now as Mount Tantalus. Lying just to the east of Honolulu, it is part of the wet, windward, Koʻolau Range, which forms a long spine that runs the length of the island of Oʻahu.

I was traveling with Brenden Holland, a biologist at Hawaiʻi Pacific University with a passion for island conservation and biogeography, and a particular interest in snails. Brenden had offered to take me up to this spot to see some snails living in a pretty unexpected environment. It was a perfect day for the trip. It had rained earlier that morning, and was still drizzling on and off: ideal weather for daytime snail spotting. When we arrived at our destination, Brenden pulled the car over onto the grass by the side of the road. We got out, accompanied by his two small dogs, and set off down the muddy trail.

It was immediately clear that this was not a "pristine" forest. All around us thrived the botanical legacies of a long history of agriculture. At least some of the area was probably initially cleared by Kanaka Maoli farmers, maka'āinana (literally, 'people that attend the land'), for the cultivation of crops. With the beginnings of commercial agriculture and plantations in the nineteenth century, numerous other species arrived. The forest around us was now dominated by crop plants gone wild: Indonesian cinnamon, coffee, guava, avocado, and the occasional, massive, old mango tree. Much of the understory of the forest, including both sides of the path, was dominated by two plants: ginger and night-blooming jasmine. It was among these two plants—both introduced in the last two centuries—that we had come to look for snails.

We had only been walking for about 20 minutes when we spotted them, right by the side of the track. *Auriculella diaphana* is not a particularly eye-catching snail, far less so than the larger tree snails of the genus *Achatinella*. In fact, this is so much the case that these tiny creatures, only a few millimeters in length with mottled brownish shells, are actually pretty hard to see. Even though it was early afternoon, in the damp weather these mostly nocturnal animals were active, as we had hoped they would be, quietly making their way along the underside of a long ginger leaf.

The presence of these snails, here in this landscape, doesn't make a lot of sense. To put the matter simply, it is not good habitat. Alongside the introduced plants, which are thought to offer less than ideal food sources, the area is also home to all of the usual predators, from rats and chameleons to wolfsnails. Despite these facts, *Auriculella diaphana* persist; in fact, they seem to be doing quite well, at least relative to many of Hawai'i's other snails. But while there are probably thousands of them left, they are now restricted to this one small spot. Drawing on surveys conducted in the early years of the twentieth century, Brenden has tried to quantify their decline. While the species was once found across an area of about 36 square kilometers, it is today found only in a single patch of about 2,000 square meters. In other words, it has been eliminated from more than 99.9% of its former range.[1] "So," Brenden explained to me as we walked, "I have a hard time calling them common. One forest fire and the species could be extinct." But for now, they hang on. Year after year, Brenden returns to find them still in this one little spot. Why they persist here and nowhere else is, in short, a mystery.

But it was actually a different puzzle that drew us into the forest that day. At the time of our visit, I was just getting to know Brenden, but I had already figured out that if I had an odd question about snails, there was a pretty good chance that he had already asked it—and perhaps even conducted an experiment or two toward an answer. It was Brenden's research on slime trail following that I drew on in the previous chapter. On this occasion, however, Brenden had agreed to meet with me to discuss gastropod movements on an altogether larger scale. I wanted to know more about how the incredible diversity of snails found in the Hawaiian Islands ended up here, how those first snails crossed vast oceans to this new home. It just so happened that this topic, snail biogeography, was one of Brenden's particular passions.[2] He told me that *Auriculella diaphana* and some of their relatives, also found on this hillside, had important insights to share. And so, we set off to see the snails in person.

Terrestrial snails are generally a sedentary bunch, not well known for their propensity to undertake long journeys. Snails tend to spend their lives close to the spot they happened to be hatched or born on (while most snails come in eggs, others are live born). When we add to this situation their very low tolerance for seawater—as their porous bodies offer no control over salt absorption and they thus dehydrate readily—the diversity and abundance of snails once found in this archipelago becomes even more confounding.

Hawai'i is made up of "oceanic islands," which means that they formed from the ocean floor, in this case over a volcanic hotspot out in the middle of the Pacific Ocean. At no stage in their history, therefore, have they been connected to any other land. As such, all of the species of plants and animals that have made their homes in this place have arrived across immense expanses of water, or evolved from an ancestor who did. This situation has left a lasting impact on the islands' animal biota. In short, animals in Hawai'i tend to have wings. When it comes to vertebrate animals, Hawai'i was for the longest time a land of birds. Until people introduced them, there were no terrestrial mammals or reptiles, with the sole exception of a winged mammal, the hoary bat.

But somehow a profusion of snails also ended up here. While these islands offer a somewhat extreme example, they are also far from unique in this regard: terrestrial snails are found on pretty much every tropical and subtropical island around the world. We are faced, then, with a kind of "snail paradox," as Brenden put it: How do organisms that are so sedentary end up being so incredibly widely dispersed?

Far from being an abstract, albeit fascinating, scientific curiosity, I am convinced that attending to snail biogeography and evolution is particularly important at our present juncture. How might these deep-time processes help us to understand the ongoing extinctions of Hawai'i's snails differently? What might this context allow us to see, appreciate, and perhaps even hold onto? As the writer Robert Macfarlane has argued, a deep time perspective can offer "a means not of escaping our troubled present, but rather of re-imagining it; countermanding its quick greeds and furies with older, slower stories of making and unmaking. At its best, a deep time awareness might help us . . . to consider what we are leaving behind for the epochs and beings that will follow us."[3]

As I walked with Brenden that day, it was these questions and possibilities that preoccupied me. But I had another interest, too. My chances to encounter snails still living in the forest have been quite few and far between. To this day, I treat each such encounter as a precious opportunity. I do so, in part, because each time I find myself in the forest with snails is another chance to listen and maybe, just maybe, hear them sing.

OCEANIC CROSSINGS

There have been surprisingly few experimental efforts to explore the possible avenues by which Hawai'i's snails might have crossed oceans to arrive in their new home. In fact, to date there has been precisely one study on this topic of which I am aware. In 2006, Brenden placed a piece of tree bark with 12 live snails of the species *Succinea caduca* into a saltwater aquarium. This is one of Hawai'i's nonendangered snail species; in fact, it is one of the few species that is found on multiple islands and seems to be doing okay. It is a coastal species, and the individuals enrolled into the study were from populations living as little as 10 meters from the beach. Brenden explained to me: "After heavy rain, they are commonly seen in gullies by the coast so there's no question that they are going to get washed down pretty frequently."

The purpose of Brenden's experiment was to determine whether, when this happens, it might be possible for these snails to move around by sea and successfully establish themselves in new places. The answer, it seems, is yes. Brenden and his colleague Rob Cowie reported that: "After 12 h of immersion, all specimens were alive, indicating that sea water is not immediately lethal and suggesting the potential for rafting between islands on logs and vegetation."[4]

Beyond Hawai'i's shores, however, there have been numerous efforts to experimentally explore or otherwise interrogate the puzzle that is the evolution and distribution of island land snails. Charles Darwin, in a letter to Alfred Russel Wallace in 1857, summed up the situation succinctly: "One of the subjects on which I have been experimentising and which cost me much trouble, is the means of distribution of all organic beings found on oceanic islands and any facts on this subject would be most gratefully received: Land-Molluscs are a great perplexity to me."[5] Or, as he put it in a letter to another correspondent a year earlier: "No facts seem to me so difficult as those connected with the dispersal of land Mollusca."[6]

In an effort to address this perplexity, Darwin submerged land snails in saltwater to discover whether and how long they might survive. Among his other findings was the fact that estivating snails of the species *Helix pomatia* recovered after 20 days in seawater. The fact that these snails were estivating is important. As we have seen, during these periods snails can create a thin layer of mucus to cover their aperture and prevent them drying out. As long as they are sealed up inside their shells in this way, it seems that many snails can survive being submerged in saltwater for weeks at a time.[7] Here we see that in yet another way, beyond the local excursions discussed in the previous chapter, slime may be a powerful enabler of snail movements.

Inspired by Darwin, a French study in the 1860s placed 100 land snails of 10 different species in a box with holes and immersed it in seawater. Roughly a quarter of the snails, from six different species, survived for 14 days—which was calculated to be about half the time it would take for an object like a log to float across the Atlantic.[8]

All of these years of submerging snails—of gastropods drowned and survived—have produced one primary, albeit tentative, finding: it is at least possible that land snails are floating around the world to establish themselves in distant places. We just don't know enough about Hawaiʻi's snails to know how likely a vector this is for their movements; we have a single, short-term study on one of the over 750 known species.

But floating is by no means the only mode of transportation open to snails. In fact, most of the biologists I spoke to were of the view that it probably isn't the primary way in which they have moved across large distances. While snails have possibly floated around *within* the Hawaiian archipelago, between islands, it is thought to be unlikely that the first snails to arrive did so in this way: the distances of open ocean are just too vast. But here, things get even stranger, and even less amendable to experimentation.

As we walked along a winding path around the summit of Puʻu ʻŌhiʻa, Brenden and I discussed some of these other potential modes of snail movement across oceans. He explained to me that not all of these possibilities are immediately obvious if we look only at organisms in their current forms. Many species change after arriving on islands; some, for example, undergo processes of "gigantism" or "dwarfism" in which their new environmental conditions lead to a significantly increased or decreased body size. Alongside these kinds of changes, as we will see, many entirely new species evolve on islands after initial arrival events. In the case of Hawaiʻi's snails, phylogenetic analysis indicates that the vast majority of species evolved in the islands in this way, a single arrival giving rise to multiple new species over a few million years (these analyses compare genetic material to determine how closely related species on different islands are to one another, and in this way piece together their histories of arrival and evolutionary divergence). Some of these new island species will continue to look a great deal like the ancestor that made that initial oceanic crossing; others will not.

It turns out that the tiny *Auriculella diaphana* we encountered there among the ginger plants are close relatives of the much larger *Achatinella* tree snails I had seen in the Palikea exclosure and elsewhere. The former is about 7 millimeters in length, the latter about 2 centimeters. But, Brenden told me, *Auriculella* and *Achatinella* have a smaller common relative still, and phylogenetic analysis indicates that it is an even more likely candidate for having made the initial trip. There, among the ginger leaves, we were lucky enough to also encounter some of these tiny beings, members of the subfamily Tornatellidinae.

The snails we saw that day, along with some other species within this subfamily, reach a maximum size of about 2 millimeters in length, roughly the size of a grain of rice. But this size difference is more significant than these simple length measurements imply. As Rob Cowie explained to me, the mass of a snail is roughly equivalent to the cube of its length. As such, one of the tiny Tornatellidinae snails might be as much as 1,000 times lighter than its *Achatinella* brethren. If a minute creature similar to these tiny snails was the ancestor that first made its way to the Hawaiian Islands, then it might have had many other modes of transportation open to it. It might even have arrived by bird.

In numerous conversations with biologists, again and again I was told with varying degrees of confidence that the most likely answer to the puzzle of Hawai'i's snails is that the first ones flew here. Everybody narrated this hypothetical scene a little differently, but the main events remained the same. At some point in the distant past, a tiny snail climbed on board a migratory bird, perhaps a golden plover, as it perched or nested overnight. As snails are nocturnal, it makes sense that they might encounter a perched bird in this way, and that this wayward passenger might then be able to hunker down, deep in the bird's feathers, sealing itself up. Days or weeks later, having rested through the exhausting crossing, the snail then climbed off the bird in its new home.

I must admit that on first hearing this explanation I was somewhat dubious. This sequence of events just seemed so horribly unlikely. I reminded myself, though, that in the vastness of evolutionary time, "horribly unlikely" is actually pretty decent odds. But as I continued to talk to scientists and read the literature, I discovered an unseen world of surprising snail journeys. For the most part, scientists have not deliberately gone looking for snails on birds, but in a handful of articles published over the last several decades they have nonetheless reported on their accidental encounters with them, usually in the course of routine bird banding or observation. In these cases, it seems, snails have sometimes been present with surprising regularity and abundance.

Across several studies, the snail *Vitrina pellucida* has been found on a variety of migratory birds in Europe, while *Succinea riisei* has been found on three different types of birds in North America, with anywhere from one to 10 snails on a single bird.[9] In one particular study, focused on migratory birds in Louisiana, snails were found on three different bird species. The main focus of the research was the woodcock, and it was only on these birds that the researchers really monitored snail presence: "Of the 96 woodcock checked, 11.4% had snails present. Of those, the average number of snails per bird was 3. . . . Snails involved in these associations ranged in size from 1.5–9.0 mm."[10] At least among smaller snails, those about the size of the Tornatellidinae, movement by bird turns out to be far from exceptional— although successful arrival and establishment in a new land, thousands of kilometers away, must still be pretty unusual.

In Hawai'i, there has never been a targeted scientific search for snails on birds, so it is hard to know which species might be climbing on board and with what kinds of frequency. Partway through my research, however, Nori Yeung at the Bishop Museum came across and shared with me a tantalizing snippet from a field notebook. The collecting note was made in 1949 by Yoshio Kondo who was at the time in Nori's current position as curator of the museum's malacology collection. There at the top of a grid-lined page, in neat cursive writing, he reported: "a juvenile sooty tern on which were *Succinea* and *Elasmias*. Brought bird back. Unfortunately, did not keep shells on bird separate."

But there is another fascinating, albeit equally speculative, avenue by which tiny snails might move around the globe. They might fly without the aid of birds, blown on leaves and other debris, or just on their own, sealed up in their shells. Indeed, there is significant evidence from sampling, conducted with nets attached to airplanes, that rock particles the size and weight of some of these tiny snails can move around in this way, sometimes being found at altitudes of more than 2,000 meters. Drawing on these findings, some scientists have argued that it is not at all unreasonable to think that snails might travel in similar ways, definitely over shorter distances but perhaps also for transoceanic journeys.[11] At least a couple of the scientists I spoke to, including Brenden and Rob, were holding open the possibility that the progenitors of at least some of Hawaiʻi's snail families may have blown to the islands in this way, perhaps even carried by the winds of a hurricane.

Of course, once a snail species has made that first giant leap across oceans, a range of other options open up for the shorter, inter-island, movements that genetic analysis indicates have taken place at various points in the past. As we have seen, some snails might survive a floating journey between islands. Others, it seems, might be making these briefer trips *inside* birds: studies in various parts of the world have now shown that a variety of snail species—including as least one species of the Tornatellidinae—can survive passage through avian digestive tracts at a relatively high frequency.[12]

These are, undoubtedly, all rather unreliable ways to travel. For every snail that successfully arrived in a strange new land on a bird or a floating branch, countless millions must have been washed, blown, or flown out to sea without such luck. The odds must be slightly better traveling by bird than log: at least in theory, if you hop onto or into a migratory bird in a forest, you are reasonably likely to be taken to another forest. Of course, for those snails unfortunate enough to be traveling inside the bird, they would have to survive the journey through the digestive system too.

While snails clearly have many things working against them, the simple fact that they are found on islands almost everywhere tells us that—despite appearances—they are actually pretty good at dispersing and establishing. While they might not fly or enjoy saltwater, they are small and robust enough to take advantage of other modes of movement.[13]

But snails have another dispersal advantage over many other animals that really becomes apparent only after arrival. As we have seen, Hawaiʻi's land snails, like many others around the world, are all thought to be hermaphrodites. This means that for successful establishment to take place, any two individuals will do. In fact, in some cases one snail might be enough: some species are capable of "selfing" (i.e., self-fertilization), and others seem to be able to store sperm from copulation for long periods of time to use at a later date. It isn't entirely clear which of these reproductive feats Hawaiʻi's many snails are up to, but there is at least some evidence of both of these possibilities among them.[14]

In other cases, even one living snail may not have been necessary. Among the diversity of Hawaiian snails are some species that lay sticky egg clusters. If their ancestors did the same, they might have traveled on a bird or a log in egg form instead. Scientists are not really sure how much of an advantage over live travel these eggs provide, especially for longer trips. In the case of snails, eggs are often prone to drying out and may not actually provide a safer, more reliable, mode of travel than a living animal sealed up within its shell.

However they travel, snails are largely at the whim of external forces in these movements, subject to what biologists call "passive dispersal." As Brenden helpfully summed it up for me: "biogeographically, snails are plants"—both groups share many of the same vectors for movement, the latter usually by seed or spore. This is clearly a "system" of island dispersal that can hope to achieve results only with immense periods of time at its disposal. Over millions of years, a few lucky snails made these journeys successfully. We can't know for certain how many times this happened in the Hawaiian Islands. But by tracing species back to their common ancestors in Hawai'i and beyond its shores, Brenden and Rob have estimated that things must have worked out for around 20, and likely fewer than 30, intrepid travelers, or groups of travelers, over roughly the past 5 million years (when Kaua'i, the oldest of the current high islands with suitable snail habitat, formed).[15] All of the rest of Hawai'i's incredible gastropod diversity is thought to have evolved in the islands from this small number of common ancestors.

While there is undoubtedly something very "passive" about this dispersal of snails—always at the whim of others, be they birds, storms, or tides, traveling under their steam and direction—this isn't the whole of the story. Deep evolutionary histories have produced these possibilities. Snails' modes of passive movement only "work" because they have evolved some remarkable traits—presumably, in this case, at least in part adaptations—for dispersal, survival, and reproduction, across and into isolated new lands: from epiphragms and sticky eggs to hermaphroditism, sperm storage, and self-fertilization. Millions of years and countless generations of more or less successful journeying have selected for those individuals that survived and established themselves best.

There is a profound kind of evolutionary agency at work here, a creative, experimental, adaptive working-out of living forms with particular capacities and propensities.[16] For the most part, individual snails are indeed relatively passive in all this. They're not, however, irrelevant. The particular actions of those snails that crawled onto a bird, that opted to seal up their apertures, that safely stored away sperm for future use, mattered profoundly. But neither are snails involved in the more active, sometimes even deliberate, dispersal undertaken by many other animals. Instead, if we pay attention, snails amaze with their capacity to move so far, to spread so widely, while doing so little. This, it seems to me, is one of the real marvels of snail biogeography. Individuals do not need to exert great effort because natural selection has acted for them, acted on them, acted with them, to produce these beings that are so unexpectedly but uniquely suited to a particular form of deep time travel, *drifting*. From such a perspective, rather than being any kind of deficiency, the highly successful passivity of snails might be seen as a remarkable evolutionary achievement.

There is so much more to learn here, so much to learn about not just the vectors but the patterns under which dispersal takes place: Are they laid down by atmospheric and oceanic currents, or by the inherited paths of avian migration? And yet to some extent this must remain a space of uncertainty and even mystery. How can one really study processes of biogeography that take place across such vast periods of time and space? As Brenden reminded me, it's likely that in the history of these islands, on average one successful snail arrival event has taken place every few hundred thousand years. Put simply, it's not something that any of us are likely to ever see, let alone study, firsthand.

As we walked on Puʻu ʻŌhiʻa that day, I heard no snails singing. Perhaps, as some biologists I asked about this phenomenon told me, this is because snails have no vocal cords and therefore cannot sing. Perhaps it was because it was daytime and snails sing only at night, when they are most active. Or, perhaps, as I noted in the introduction to this book, it was because snails sing as a sign that all is "pono"—all is righteous, correct, and good—and everything is a long, long way from being right in the world, especially in the worlds of snails.

But much else about this singing also remains elusive, at least to me. I'm not sure, for example, what snail songs sound like. They are described by some as beautiful; others told me they chirp, or that their song takes the form of a single, high-pitched note. There is also some uncertainty about which snails sing. Many people now associate singing only with the larger tree snails, like those of the genus *Achatinella* and perhaps their close relatives of the genus *Partulina* found on several islands including Maui, Moloka'i, O'ahu, and Hawai'i. While the mountain we were walking on was once home to *Achatinella* tree snails, they are now long gone. So, perhaps this was not even the right forest to be listening in. In the historical records, though, names like pūpū-kani-oe ('shell sounding long') and pūpū-kani-ao ('shell that sings at dawn'), as well as accounts of singing snails, extend to all of the islands, and are sometimes also explicitly associated with other species, such as the large ground-dwelling *Carelia* species once found on Kaua'i.[17] So, then again, perhaps there might have been a chance to hear singing here after all.

With so much left to learn, my approach has been to keep an ear open in all places and to all snails, even the tiny ones like *Auriculella diaphana*, just in case. I also took every opportunity to ask people about this singing. In almost all cases, I received one of two kinds of replies.

For many people, the most logical explanation for these stories was a case of mistaken identity. Alongside their many snail species, Hawai'i's forests are also home to a diverse range of crickets, some of which chirp loudly, sometimes beautifully, at night. They're also very good at keeping quiet when someone approaches, and at jumping out of the way, remaining much more inconspicuous than many of the larger snails. This explanation for the song of the snail seems to have been first proposed by Robert Cyril Layton Perkins, the "Father of Hawaiian Zoology," in the late nineteenth or early twentieth century, and has been widely accepted by many biologists since.[18]

In their published discussion of Kanaka Maoli stories of singing snails, Aimee You Sato, Melissa Renae Price, and Mehana Blaich Vaughan offer some important context for this attribution of song, noting that it may have been reinforced by cultural predispositions. As they put it, drawing on in-depth discussions with a range of cultural practitioners: "In the Hawaiian mind, a beautiful voice would naturally belong to something as exquisite as the kāhuli."[19]

But other people I spoke to in Hawaiʻi found the mistaken identity explanation highly unlikely, even a bit insulting. Kanaka culture, they noted, is rooted in an intimate and intricate knowledge of the ʻāina and its community of life (discussed further in chapter 3). How could these people possibly have made this error?

Instead, many of these people offered an alternative account, telling me that stories of singing snails refer to the whistling sound that might be made as the wind rushes over the aperture, or opening, of a snail's shell as it hangs from a leaf or branch. Some scientists told me this was unlikely. According to Mike Hadfield, these shells were just too small to produce sound like this. What's more, Mike said, snails tend to assiduously keep their apertures covered: "no self-respecting snail would give up its moisture by sitting there gaping into the wind."

But this explanation remains a popular one. It is thought by many of the people I spoke to that in centuries past, when the forest was thick with snails, this sound would have combined to produce a melody of sorts. This connection between wind and singing is also noted in some historical sources. The chant "Pa Ka Makani" tells us that "Pa ka makani, haʻu ka waha o ke kāhuli i ka nahele / The wind blows, the land shells trill (mouths tremble) in the forest."[20] These days, however, much has changed in these forest places. Alongside the incredible decline in snail numbers, and so potentially whistling shells, these places have experienced large-scale degradation, from deforestation to introduced species. People argue that as a result of these changes, as Sato and her co-authors put it: "the wind does not travel through the forest in the same way as in the past."[21] From some of the people I spoke to I had the sense that these changes might have brought about an end to the snail's song, at least for now. Others, however, were clearly of the view that, in the right conditions, with some care, the voice of the snail might still be heard.

About halfway through our walk that day Brenden and I stopped in a thick patch of ginger and jasmine to look for more snails. When we turned over the first leaves, sure enough, there they were. And on the next one, too. In fact, they were on almost every plant in this particular area. *Auriculella diaphana* certainly seemed to be doing well in some patches.

But perhaps not everything is as it seems. Just a few years ago, Brenden and two coauthors—Luciano Chiaverano and Cierra Howard—took some of these snails into the lab to explore what impact their altered habitat might be having on them. The diet of *Auriculella diaphana* consists of the thin layer of fungi and other microbes that they scrape from the surface of leaves. Brenden and his collaborators wondered if the different plant species might host different microbial communities in a way that matters for snail health.

For periods of 4 months, the researchers housed snails in one of three situations: with common native plants, with the nonnative ginger and jasmine, or with a mixture of the two. While all of the snails in all of the treatments survived, the differences in their fecundity were dramatic. The snails housed with native vegetation produced 20 times the number of eggs of those that only had nonnative vegetation, and roughly 15 times the number of those that had a combination of the two plant types. "These results," they argued, "suggest that native snails persisting in nonnative host plants experience sublethal stress, reflected in a dramatic reduction in reproductive output."[22]

This stress is a form of what the literary scholar Rob Nixon has called "slow violence."[23] Alongside the more dramatic and visible violence of habitat destruction that has impacted on many snail species, and their outright mortality through predation or collection, here we see a drawn-out process in which the conditions for ongoing life are quietly undermined. The lush green sea of vegetation simply does not nourish as it once did.

It remains unclear why this is the case. As the researchers put it: "The mechanisms underlying enhanced reproductive output observed in native plant cages in this study are unknown." It seems likely, however, that nutritional differences between the available microbes are at least part of the story.

The forest habitats that Hawai'i's snails co-evolved with have been being dramatically altered for many centuries now. Each set of changes has brought fresh challenges. Especially in lowland forests, agriculture has been displacing snails since Polynesian peoples first arrived in the islands. As Mike put it in a published paper, the "clearing of lowland vegetation was probably responsible for unrecorded extinctions of low altitude populations or even species."[24] Those species with relatively small ranges would have been particularly hard hit.

At the same time, the arrival of Polynesian rats on board wa'a (voyaging canoes) during this period would have impacted significantly on snails. In the past, these impacts were thought to have primarily taken the form of direct predation—rats of all kinds are dedicated snail predators—but in recent years a growing body of paleobotanical work in Hawai'i and around the Pacific has shown that rats also transformed forest ecosystems in previously unimagined ways through their voracious appetite for larger seeds and fruits.[25] Drawing on this research, biologists Sam 'Ohu Gon and Kawika Winter have argued that consumption by rats of the large-seeded plant species that formerly dominated Hawai'i's forests "induced an ecological shift toward a forest dominated by smaller-seeded species. This regime shift resulted in a series of cascading extinctions that swept land crabs, land snails, and flightless land birds out of existence."[26]

But the widespread destruction of Hawai'i's forests accelerated exponentially from the early nineteenth century with the arrival of European explorers, traders, and whalers, and the subsequent opening of the islands to global markets and a range of newly arrived species. One of the most significant impacts on Hawai'i's forests in this period resulted from the boom in sandalwood trade to China under King Kamehameha I. Kamehameha (1736–1819) was an ali'i (chief) from the island of Hawai'i who rose to power to unify the whole archipelago under his rule, by both conquest and treaty. A significant part of Kamehameha's military power lay in his access to European munitions, munitions that were purchased in part by the proceeds from a monopoly on the lucrative export market for sandalwood. Kamehameha dominated this trade until his death in 1819; a decade later, these trees were all but exhausted.[27]

During this same period, a second major threat to Hawaiʻi's forests was introduced to the islands in the form of cattle and other livestock. The British Captain George Vancouver first brought cattle to the island of Hawaiʻi in the early 1790s, offering them as a gift to Kamehameha. The animals promptly multiplied, protected by chiefly privilege.[28] Cattle spread out across that island, and a few decades later were also to be commonly found in the forests of Maui and Kauaʻi.[29] Eventually, they were established on all of the main Hawaiian Islands. As they went, they destroyed the forests, eating or stamping out any new growth so that, as one observer writing in the 1850s reported: "whilst the old trees die of age, no young ones are seen taking their places."[30]

These cattle also severely damaged the crops of the makaʻāinana, who were prohibited from interfering with their movements, initially because of the chiefs and later because of the power and influence of wealthy ranch owners. Indeed, the refusal to control cattle, or to allow others to do so, is an important strand of the story of how Kānaka Maoli were eventually displaced from their traditional lands in Hawaiʻi (a topic that is taken up in more detail in the next chapter). Across the islands, large ranches frequently deployed a strategy of "buying out kuleana [small land parcel] owners who had grown weary of battling the cattle."[31] This is a process that has been observed in different ways in many parts of the world, livestock serving as shock troops of empire.[32]

During the period of early contact in the late eighteenth and early nineteenth centuries, outsiders also brought with them a range of devastating diseases such as gonorrhea, syphilis, and tuberculosis which spread rapidly through the Hawaiian population, who had little immunity. The death toll was staggering. According to some accounts, primarily as a result of disease, by 1850 the population of the Hawaiian Islands had been reduced by as much as 90 percent. Disease and death tore at the fabric of Kanaka society, impacting families and political structures, leading to chronic labor shortages, as well as causing "reduced life spans, rising infertility and infant mortality, and persistent poor health for generations of Islanders."[33]

This social upheaval, of course, rippled out into the wider Hawaiian landscape. The island's ecosystems began to be significantly transformed by the breakdown of agricultural and irrigation systems as the population declined. This situation paved the way for the rapid transformation, over just a few decades, from a subsistence economy to one grounded in industrial agricultural exports.

As historian Carol A. MacLennan has argued, sugar is key to the story of Hawaiʻi's environmental transformation: "The first Western-style sugar plantations appeared amid Hawaiian communities in the 1840s. Plantation agriculture took off in the 1880s with large infusions of capital. By 1920, sugar had remodeled the islands into a production machine that drew extensively on island soils, forests, waters, and its island residents to satisfy North America's sugar craving."[34] Vast areas of land were taken over for sugar production during this period. More than this, though, MacLennan emphasizes the way in which plantations spread out into the Hawaiian landscape, transforming everything as they went. They did so in part through their "subsidiary industries of ranching and rice growing," but equally as importantly through the way in which much of the remaining forest was reimagined as a water catchment whose primary utility was in providing water for this thirsty crop.[35] Meanwhile, in the early decades of the twentieth century, other significant areas of land were being converted by many of the same companies for pineapple plantations.[36]

These are the key processes that wiped out or seriously degraded much of the forest habitat on which Hawaiʻi's snails depended. In the area of Puʻu ʻŌhiʻa where Brenden and I were walking, all of these forces came together. By the turn of the twentieth century, the impacts of land clearing and logging, as well as those of cattle and other ungulates, "had laid bare vast areas of hillsides."[37] In an effort to combat this loss of forest, both here and throughout the islands, a massive reforestation program was launched in the first decades of that century. This work aimed, in particular, to restore and protect the islands' many degraded water catchments. In place of the native forest, fast-growing tree species like macadamia, mango, jackfruit, eucalyptus, and camphor laurel were all planted. Much of the eclectic forest found on this hillside today no doubt has its roots in these efforts. In fact, this place was one of the central hubs of this activity: in the late nineteenth century, 13 acres of forest still remaining in this area was cleared to make way for an experimental station to trial many of these plant species.[38]

Standing among the incredible variety of plants on that hillside, produced by this history of land use changes, it is not hard to understand why scholars like Anna Tsing and Donna Haraway have suggested that an apt name for our current period of environmental transformation might be the "Plantationocene."[39] From this perspective, it is not so much an amorphous humanity—the Anthropos of the Anthropocene—that is driving the transformation of our earth, but particular modes of human life, ways of relating to the land, to crops, and to the often destructive possibilities for generating wealth that lie within them. All over the world these transformations have been bound up with and enabled by diverse forms of exploitation, dispossession, and violence toward particular human bodies and communities. In Hawai'i, the plantation and its close cousin, the ranch, have had just such an impact, remaking life in these islands for everyone, from people to snails. While the heyday of sugar and cattle ended long ago on O'ahu, their legacies continue to be felt, to be lived and died with.

Through this ongoing change to Hawai'i's forests, the intimate relationships between snails, plants, and the broader environment have been degraded, transformed, or simply destroyed. There is so much about these relationships that we don't really understand. What role, for example, did huge numbers of tree snails once have in the environment? Some preliminary work by Brenden, Richard O'Rorke, and others has suggested that leaf-cleaning might have helped to ensure a high level of diversity of microbial communities that may in turn have limited the spread of pathogenic surface microbes that harmed plants.[40] Other people have suggested that regular cleaning by snails may also have enhanced plants' capacities for photosynthesis.

The numerous species of ground-dwelling snails may also have played important roles in the forest, breaking down leaf matter and other organic materials to produce healthy soils.[41] In fact, until earthworms and other detritivores were introduced to the islands in recent centuries, this important job is thought to have fallen to the snails. Perhaps this contribution was a particularly important one.

It's also possible that these snails would have fulfilled the same glamorous ecological role that they do in most other parts of the world: being eaten by others. In Hawai'i, large numbers of snails might have been a significant source of food for some animals. However, until the relatively recent arrival of their current predators, there weren't many animals in the islands that are good candidates for having eaten snails, especially the bigger ones. Some people have speculated that in the distant past, birds now long extinct—including large flightless ibises and rails—may have eaten snails, perhaps in significant numbers.[42] The only Hawaiian bird species to have been documented eating snails is the po'ouli (*Melamprosops phaeosoma*), a little forest bird that was described by science only in 1973. Sadly, the last po'ouli was sighted in 2004, but in this brief window of time between discovery and extinction these birds were frequently seen foraging among the bark and moss for snails.[43]

Ultimately, however, we have no way of really knowing just how significant these ecological roles were prior to the decimation of snails, birds, and their forests in Hawai'i. There is nonetheless a strong desire among conservationists to have this kind of ecological story to tell about the importance of the species they are working to protect. Such a framing taps into deeply entrenched notions of how ecosystems function, as well as established understandings of how to appeal to publics in the name of conservation.[44] As Aldo Leopold, one of the founding figures of US environmental thought, put it many decades ago now: "to save every cog and wheel is the first precaution of intelligent tinkering."[45]

While I am not enamored of these machinic metaphors, there is definitely something to this position. But in the case of Hawai'i's snails, if they are or were a cog, it is largely unclear what function that cog played in the larger system. Those rare parts of the forest that still contain many snails don't seem to be any healthier than the vast majority that is now devoid of their presence. If ground-dwelling snails did once play a role in soil health, they no longer seem to be needed in this way, likewise for the disappeared birds that may once have consumed them.

While a simple functional story about the current ecological importance of snails in the forest might be convenient and helpful, the reality is that we just don't have one. If it exists at all, it is elusive—made all the more so by the large-scale destruction of Hawai'i's forests. And yet, as Mike put it when I raised this topic with him: "anything that was as abundant as the snails were in the forests of 1850 was an important part of the ecosystem. They were eating a big part of it, pooping a big part of it." While we cannot name these relationships as precisely or definitively as we might like, we have to imagine that they existed in some form, and might still, albeit differently. For Mike, it simply must be the case that with the loss of these snails, "whole elements of connectivity are gone."

While we may not comprehend them in detail, the very particular relationships that snails have had with plants, soils, and landscapes are key to understanding how it is that Hawai'i has ended up with such a spectacular diversity of terrestrial gastropods. As we have seen, the vast majority of these species are thought to have evolved in the islands. Single arrival events gave rise to one new species, and then another, and then another, as populations separated and diverged. In no small way, it was the Hawaiian environment that produced the conditions for this radiation of diversity. It did so in part through the relatively ideal environment that it provided: plenty of moist, entirely or largely predator-free forests. These environments offered gastropods a range of adaptive possibilities. But perhaps more importantly, the Hawaiian Islands also provided ample opportunity for the kind of separation and isolation that allows snail populations to simply drift apart and become different species.

The main Hawaiian Islands, the high islands, are lands of significant ecological diversity and patchiness with rainfall and vegetation often varying dramatically across relatively short distances. While many people living beyond their shores associate these islands with a mild, unchanging, climate, the reality is anything but. "Despite their small total area, the Islands have places which are desert dry and others which may be the wettest places on earth. Temperatures can go below freezing on high mountains, or can be above 90°F at sea level. Humidity can stand at virtually zero in alpine areas, yet be 100% nearly all of the time in wet mountain areas."[46]

Many of these landscape features form impassable barriers for snails, at least most of the time. Unlike birds and other highly mobile animals that can spread across an island into all areas of similar, suitable habitat, snails rely on occasional dispersal across barriers by intermittent forces such as storms, hurricanes, floods, slope failures—and perhaps hitching the occasional ride on a bird. This means that snail populations, once separated, are less likely to reconnect over extremely long intervals. A key study conducted by Brenden and Mike bears this out: they found that populations of *Achatinella mustelina* separated by only a few kilometers had experienced no significant gene flow between them for hundreds of thousands, perhaps even a million or more, years.[47] This kind of isolation creates ideal conditions for populations to simply drift off in different evolutionary directions, giving rise to one or more new species in each locale.

In other cases, movements of the land itself might open up new barriers. This could happen through a landslide that carries some snails out to sea and over to another island, or it may happen over immense periods of time as a deep valley is eroded out of a volcanic mountain chain, creating a warmer and drier barrier zone that is less hospitable for snails between the two peaks.[48]

Over the long duration of evolutionary time, there are numerous paths to this kind of separation of one or a few snails from a larger population. While the landscape of the Hawaiian Islands offered an ideal evolutionary environment, snails, too, played their part. Snails are ideally suited to this kind of speciation as a result of their largely sedentary form of life that is punctuated by occasional moments of long-distance dispersal. In this unique combination of landscape and organism, there is a perfect recipe for speciation. As a result of this situation, the vast majority of Hawai'i's snails are what biologists call "single island endemics," restricted to an individual island, and in fact are often found only within very limited ranges on any given island.

As David D. Baldwin (1831–1912), one of the first naturalists to turn his attention to these snails, put it in his discussion of the mountains of Oʻahu in 1887: "The sides of these ranges the entire length are furrowed with deep valleys separating lofty ridges. These valleys and ridges are the home of the Achatinella. Each valley and ridge has its own distinct species which are connected with those of the next valley and ridge by a multitude of intermediate varieties presenting minute gradations of form and color."[49]

Importantly, however, the ecological and evolutionary stories offered by these snails are not simply those in which gastropods "adapted" to, or were otherwise altered by, an existing forest landscape. We often slip into telling evolutionary stories in this way: imagining a relatively static, pre-given environment that creates the conditions that species respond to over evolutionary time.[50] While such a perspective has faced sustained criticism, both long before and ever since Darwin, its simplistic logic continues to rear its head.[51] In reality, any given ecosystem is itself shaped in many ways by the species that comprise it—which are themselves evolving. Evolution is a multidirectional process of *coshaping*.

From such a perspective we must also acknowledge the ways in which Hawai'i's snails, over millions of years, helped to produce and maintain these forest ecosystems, rather than just being shaped in or by them. This understanding begins to unsettle the notion that evolution is something that simply "happens" to a species.[52] It opens up space for an appreciation of the many ways in which the lives, relationships, and decisions of living beings help to shape how they and others *become*, in part through the ways in which they can influence the environments and other species that they are in turn shaped by. Again, there remains much to be learned here about Hawai'i's snails, and much that will probably never fully be understood.

It should perhaps come as no surprise that studies of Hawai'i's snails played a role in the development of scientific understandings of what evolutionary biologists now refer to as "genetic drift," a process of nonadaptive radiation. The work of the nineteenth-century naturalist and missionary John Thomas Gulick (1832–1923) was particularly significant in this regard. Deeply impressed by Darwin's work on natural selection, then only recently published, Gulick set out to explain how the amazing diversity among the *Achatinella* tree snails he saw around him on the island of O'ahu might have evolved.

Gulick noted, like his contemporary Baldwin, that within forest environments that seemed to be pretty much identical to one another, he encountered snails with significantly different colors and patterns.[53] Why, facing almost exactly the same conditions, and therefore selection pressures, did so much variation arise? Could it really all be adaptive? Gulick didn't think so. In dialogue with George John Romanes and other evolutionary theorists of the day, he proposed that there was an "inherent tendency to variation" within organisms. As a result, isolated populations might slowly diverge from one another without any meaningful differences in their environments.[54]

While this perspective faced serious opposition at the time, today this kind of genetic drift is widely accepted, alongside adaptive selection, as a key aspect of evolutionary change. In their own small but important way, Hawaiʻi's snails—emerging out of their very particular history of arrival and change in this place—played a role in crafting contemporary understandings of the evolutionary processes that shape our living world.

Telling these kinds of snail stories requires us to think carefully about how we understand and talk about island places and their relationships with a broader oceanic region. Decades of Indigenous scholarship from around Oceania has pushed back against the understanding of islands as tiny, remote, patches of land in a vast ocean. Scholars like Epeli Hauʻofa have emphasised that these ideas about the region are grounded in a not-so-subtle continental thinking. As he noted in a now classic essay:

> There is a gulf of difference between viewing the Pacific as "islands in a far sea" and as "a sea of islands." The first emphasises dry surfaces in a vast ocean far from the centres of power. When you focus this way you stress the smallness and remoteness of the islands. The second is a more holistic perspective in which things are seen in the totality of their relationships.[55]

Focusing on these relationships yields a vision of Oceania as an interconnected region, one in which islands and their peoples have long been in relationships, sharing histories, ideas, resources, and much more. Of course, many of those relationships were severed or degraded by processes of colonization in the region. As Tracey Banivanua Mar has argued, colonization "redrew the boundaries and borders that had previously joined and separated peoples from one another"—in some cases, connecting them in new ways through globalized networks, but frequently creating new modes of disconnection through various forms of "imposed isolation."[56] More than an abstract question of academic interest, many Pacific scholars have emphasized that this notion of the region as comprised of small, isolated, and poorly resourced patches of land, has been increasingly internalized by local people to, in Hauʻofaʻs words "inflict lasting damage on people's image of themselves."[57]

Importantly, a related mode of thinking about this region and about islands more generally also dominates in the biological sciences. We see it in the way that island ecosystems are frequently described through comparison with an assumed continental norm. From this perspective, islands are presented as having "disharmonic" distributions that are disproportionately rich or poor in terms of the diversity of certain groups of plants and animals. One of the key differences frequently noted in this regard is a relative lack of predators for some species on many islands, leading to adaptations— like flightlessness among birds—that are less common on continental landmasses. These kinds of adaptations, combined with the relative isolation of islands, have grounded a general assumption that these places are "evolutionary dead ends." In short, any species that do arrive and establish themselves are never going to make it back to a continent—or at least close enough to never for it not to really be worth looking for. In recent decades, however, this understanding has been increasingly unsettled, with a range of plants and animals being found to have made such journeys.[58] Tellingly, these modes of understanding islands—especially as isolated and contained systems that might be studied (island laboratories)—played a role in enabling the use of these places in the Pacific and elsewhere as sites for nuclear weapons testing, a process that was itself thoroughly bound up with the emergence of the "ecosystem concept."[59]

These are some of the complex threads of understanding that snails lead us to consider. As we have seen, the particularities of this region, its oceans and its landmasses, have profoundly shaped the diversity of snail life that has emerged in Hawaiʻi. The size of these islands matters, as does their ecological patchiness, their mountainous terrain, and their capacity to support moist forests. Their distance from other snail-rich places also matters, influencing the (in)frequency of new arrivals and the vectors by which they might make such journeys. But we have also seen that this is anything but a simple story of small, isolated, places. Perhaps most importantly, these particulars of size and distance are not diminishing attributes: rather, in Hawai'i they have been an incredible engine for the diversification of snail life, and the evolution of one of the richest assemblages of gastropods found anywhere on the planet. At the same time, attending to snails reminds us of yet one more way in which this sea of islands has always already been profoundly connected. Places that may have seemed to some of us to be distant and isolated, have actually been shaped, right down to their snails, by ongoing patterns of movement and exchange that are utterly foundational to the environments we all inhabit.

ON MYSTERY

As Brenden and I got ready to head home, we reached a fence. Opening the small gate and stepping through, we entered into a different landscape. About a decade of work by the volunteers of the Mānoa Cliff Native Forest Restoration Project had transformed this area. Manual removal of some of the more invasive plants, along with a fence to keep pigs out, has allowed native plantings to again take hold.[60] All around us, koa, mamaki, and other Hawaiian plants species were thriving. But there was not a snail in sight.

Brenden and others have searched the area for *Auriculella diaphana*, but to no avail. A mere 15 minutes' walk down the track, and the snails can be found everywhere. But here, among vegetation that could offer them a better chance at a future, they are nowhere to be found. Brenden wonders whether translocating snails into the area might be a good idea, but it is a risky prospect. There are just so many unknowns. These species of snails were wiped out from the area once before. Perhaps the replanted vegetation would give them a better chance of survival. Or perhaps, as Brenden noted in a published article, the nonnative vegetation where they're currently living "offers some as-yet-unrecognized form of predator protection."[61] No one can really say for sure what is going on here, why these snails persist in the particular, seemingly unlikely, places that they do.

Standing in this little fenced patch of forest it was hard to fail to appreciate how profoundly fractured and isolated the lives and possibilities of Hawai'i's snails have become. The sad reality is that many of these species can now survive only in tiny, sequestered patches of space. For the most part these are areas of active exclusion and containment, like the exclosures and the environmental chambers in the laboratory. Now, here I was behind a pig fence, wondering if this little area might also become some sort of safe haven for a population that has itself already become an isolated remnant, albeit one that endures for unknown reasons.

In short, in Hawai'i it seems that increasingly snails can only survive cut off from the environments, processes, and relationships that birthed them. Environments that once provided the conditions not only for survival but for an incredible radiation of diversity have become lethal or, at best, sublethal.

While reflecting on snails and fences that afternoon, with the air filled with the sounds of birds and crickets, my mind wandered to a conversation I had had earlier that week with Puakea Nogelmeier, an expert in Hawaiian language and culture. When I asked Puakea about the traditional stories and ideas about snails singing, he replied with a story of his own. Many years earlier, Auntie Edith Kanaka'ole, the renowned composer and kumu hula, told him and a group of her chant students that scientists had taken her to their lab to explain how impossible it was, biologically, for a snail to sing. He continued: "Auntie Edith's take on that was: 'Isn't that sad, they won't sing for the scientists.'"

Auntie Edith's understanding draws us into another space of uncertainty, or mystery, in the unfolding story of these snails. Uncertainty and mystery are not the same thing. While the former describes a lack of knowledge that might at some stage be acquired, the philosopher David E. Cooper tells us that mystery—at least in the strong sense of the term—is an acknowledgment that the world is not, even in principle, fully *knowable*.[62] Likewise, in her response, Auntie Edith reminds us that the world, and the other living beings that comprise it, are not objects transparent to our gaze, readily revealed. Each of us—scientist, kumu hula, philosopher, snail—knows one another, knows the world, from the *inside*. Which is to say that each of us only ever knows the world in a partial way; much will always remain, *must* always remain, unknowable, or simply knowable in other ways.

The anthropologist Deborah Bird Rose argued that this kind of mystery is an inherent and unavoidable feature of life. As she put it: "One cannot remove one's self from the system under examination, and because one is a part of the system the whole remains outside the possibility of one's comprehension." But, Rose insisted, this mystery is a cause for celebration; it is something to be respected, cherished, even guarded. Mystery signals the complexity and integrity of our world. A system that is entirely knowable is a dead or dying one: "total predictability would signal crisis—loss of connection."[63]

From this perspective, mystery is inseparable from the vitality of the living world. This is by no means a celebration of ignorance, but rather a humble acknowledgment of the many layers and possibilities of our shared world. It might perhaps be seen as an acknowledgment of what is called kaona in ʻōlelo Hawaiʻi (the Hawaiian language), a term that is often used to refer to the many meanings—some of them hidden—carried within a text, but that, as Kanaka historian Noelani Arista notes, might also describe the more general "tolerance and preference" in Hawaiian thought "for multiplicity in the relation between not only words but also *worlds*."[64]

For as long as they or we endure, Hawaiʻi's snails will continue to draw us into uncertainty and indeed mystery. These spaces of not knowing might reside in the one or many vectors by which these snails arrived in the islands, in relation to their possible ecological roles as cleaners and decomposers of leaves, or perhaps in relation to the ways in which they might sing and make meaning in the world with and for the peoples of Hawaiʻi. Today new mysteries are being added into the fold: Why do snails persist here and not there, seemingly against all odds? Where might their best chances of future life lie?

As Hawaiʻi's snails disappear, the vitality of these mysteries is threatened too. While these diverse spaces of not-knowing cannot really be lost in any absolute sense, the possibility of our *living well* with them can be: the possibility of inhabiting the world in a spirit of respectful curiosity, humility, and wonder.

As species slip away and the lives of those that manage to survive become increasingly fractured and simplified, we undermine our ability to find ways to relate to and value a living planet that is much bigger and more complex than we can know. We undermine the possibility of looking at a snail and seeing a mystery that can never be uncoiled but that must rather, simply, be lived with. Questioned, explored, wondered about, chanted, danced, and sung with, but ultimately always to remain at least partly unknown.

Snails are just one way, or rather one constellation of ways, into a world that exceeds us. None of the snails I have encountered in Hawai'i has sung in my presence; or if they did, I was not able to hear them. But thinking carefully with and about snails nonetheless drew me into new understandings, new modes of appreciation and respect. For me at least, the immense biogeographical and evolutionary stories coiled within their tiny shells provide a portal into processes of incredible scope and complexity, of exciting understanding alongside ongoing and profound mystery.

It is hard to really make sense of the vast, deep-time assemblage of Hawaiian snail life. I imagine it as something like a giant network with strands stretching out across the Pacific Ocean and beyond, extending back over evolutionary and geological time frames. Each strand represents one of hundreds of unique species. Millions of years of unlikely journeys—nestled into a bird's feathers, or perhaps tucked away in the crevice of a floating log— heading to destinations unknown. Millions of years that have produced these intrepid, even if somewhat unlikely, island dispersers with the reproductive and other adaptations that made these movements possible. And then, after that most fortuitous of arrivals, countless generations more of chance movements giving rise to isolation and speciation.

Drifting snails, drifting valleys and hills, drifting genes; each strand is a unique line of movement, of relationships, of branching transformations. These are at least some of the processes that have produced the breathtakingly diverse, utterly unrepeatable assemblage of snail life in Hawai'i.

To labor to hold this network in mind, however imperfectly, however impossibly, might offer us another glimpse into why these snails matter, and so the significance of what is being lost in their extinction. Doing so might remind us that each of the fragile, fleshy, little individuals of *Auriculella diaphana* or *Achatinella mustelina* surviving on this exposed mountainside or tucked away in the Palikea exclosure is not so much a "member" of a species as it is a "participant" in a lineage, one link in a vast, improbable, intergenerational project. Each of these specific projects is a shifting and emergent one. Through processes of adaptation, of isolation, of drift, a species is always becoming other than itself: changing, diversifying, multiplying. As such, what is lost with each species is not "just" its current form—not just the particular mode of life and physical manifestation of these snails we see around us today—but all of what they have been, and all of what in the fullness of evolutionary time they might have become, and might have enabled others to become with them.[65]

In Hawai'i today, the lives, histories, and possibilities of diverse snail species are being radically truncated, or simply shorn off, all within the space of a few generations of human life. With them is disappearing countless unique ways of life and the vast evolutionary heritage—to borrow Loren Eiseley's apt term, the "immense journey"—that they together comprise. But with them is also disappearing the possibility of our learning to be at home in the world in a way that is responsible for other modes of life: not in isolation, not as barely surviving populations and organisms, but as ongoing processes of entangled life and death stretching across oceans of time and space that we cannot comprehend.

3 THE COLLECTED

The Ongoing Colonial Entanglements of Shells

The clouds quickly disappeared from sight, as if being chased out of the valley. The rain that had been falling steadily for the last couple of hours had only just let up, but in a way that I was coming to understand to be characteristic of the changeable mountain weather of O'ahu, a stormy day had become another beautifully sunny one. I was in the mountainous back section of Ha'ikū Valley, on the eastern, windward side of the island. All around me, sheer, jagged cliffs rose up, containing an amphitheater-like valley of lush green.

Cutting its way through this incredible landscape is the ever-present form of the Interstate H-3, the gentle rumble of traffic accompanied by the visual spectacle of a four-lane highway suspended about 15 meters above the trees on massive concrete pilons. The construction of this highway—said by some to be the most expensive interstate ever built, and to have been quietly funded by the military in an effort to connect up key bases—caused significant controversy. Concerned activists pointed to impacts both on the environment and on Kanaka Maoli cultural sites. This opposition significantly delayed the highway's construction—which ended up taking over 30 years—but was ultimately unable to prevent it.[1] The highway opened in 1997, and to this day some Kanaka families still refuse to use it.

The Interstate H-3 also holds a particular significance in the recent history of the island's snails. It was during a period of strong community opposition to its construction, in the early 1980s, that the *Achatinella* tree snails were listed under the US Endangered Species Act (ESA). But this listing did not progress smoothly, and this highway might be part of the reason why.

In January 1981, compelled by the evidence presented by Alan Hart, Mike Hadfield, and others, the United States Fish and Wildlife Service (USFWS) published a Final Rule in the Federal Register declaring "all species of the genus *Achatinella* to be Endangered."[2] In keeping with standard practice, however, the rule was published with an effective date 30 days into the future. In most cases this wouldn't make much of a difference, but only a week after publication Ronald Reagan became the 40th President of the United States. Two days later, Reagan announced the establishment of a Presidential Task Force on Regulatory Relief, which aimed to, in his words, "cut away the thicket of irrational and senseless regulations . . . and to set loose again the ingenuity and energy of the American people."[3]

The ESA was a particular target of this process, and the Reagan administration suspended all pending additions to the list. The snails, along with three rare plants—two from New Mexico and one from Texas—had their listing deferred on a rolling month-by-month basis, pending further review.[4] While these particular species were directly caught up in this process, it was clearly part of a larger agenda in which the administration tried "to convince Congress to back away from the terms of the Act," allowing more consideration of the potential economic impacts of listing and conserving endangered species.[5] Or, failing that, as one commentator put it a few years later, to slow the listing process "to the pace of an endangered Oʻahu tree snail."[6]

In Mike's view, however, the *Achatinella* had a particular significance in these federal machinations. In early 1981, he and colleagues had accepted a contract to survey the area of the proposed Interstate H-3 for these snails under the assumption that their listing was imminent. While their survey, conducted from mid-March to mid-April, found many shells from extinct populations, no surviving snails were encountered. Shortly after submitting their report, the listing of the snails was allowed to formally go through. All of the remaining members of this genus achieved endangered species status on August 31, 1981. Had living *Achatinella* been found in the area—pesky snails that might become a further obstacle to the completion of this strategic highway—Mike suspects that things might have worked out very differently.

My reason for visiting Haʻikū Valley on this day was a bit less snail-focused. I was here to meet with Cody Pueo Pata, a distinguished Kumu Hula. Pueo had asked me to meet him at his work, an educational center called Papahana Kuaola that occupies a 62-acre area toward the back of this storied valley. Here, a team of staff and volunteers work with students of all ages, teaching them to draw on the rich cultural heritage of these islands—from farming and resource management to the knowledge embedded in traditional moʻolelo (stories) and hula—to contribute to pressing social and environmental challenges. In the words of the organization's website, they are "an ʻāina based learning organization that is connecting the area's past with a sustainable future."[7] The Hawaiian word ʻāina is often translated as "land"; alongside the earth itself, the term refers to the living community that inhabits it, understood as the provider and source of nourishment, the cultural context of all life.

It was ʻāina that I had come here to speak with Pueo about. Of course, I hoped that he might be willing to share some snail stories and insights with me, but my more general purpose in contacting him was to learn about the relationship between Hawaiian culture and the specific landscapes, plants, and animals of these islands.

As we sat and talked, Pueo described to me the work that they carry out within his hālau hula (school of hula), called Hālau Hula ʻo Ka Malama Mahilani. This is a work that is a world away from the iconic postcard images more familiar to most people raised outside the Hawaiian Islands, in which hula is generally portrayed as a frivolous activity focused on entertaining tourists.[8] As Pueo explained to me, hula is a rich cultural practice, one that is grounded in an intimate knowledge of and relationship with the ʻāina.

At the heart of the hula practiced in this hālau is the kuahu, the altar. Drawing on his more traditional, classical lineage of hula training, Pueo explained that "the kuahu is a place where we make ready for the akua, the different deities, we entreat. We call the akua to come sit within the kuahu." All of the activity of the hālau hula centers on the kuahu; the dedicated performance of the hula dancers, their skill, their commitment, their sweat, and their mana (energy) are directed toward this focal site in which the akua can be addressed, acknowledged, and honored.

This work, Pueo explained to me, requires hula practitioners "to crawl in the forest from the top to the bottom . . . to see the entire breadth of the ahupuaʻa" (a traditional land division that often runs from the mountains to the sea). The tops of the mountains have always been sacred places to Kānaka Maoli: this is Wao Akua, the realm of the gods. These are the places that catch the clouds and bring life-giving fresh water to the islands. For the hula practitioner, trips into the forest are frequently focused on collecting plants. Each of these natural entities is understood to be a kino lau, a body form or manifestation of a particular akua.[9] They must be collected following the proper protocols so that they might be placed onto the kuahu or woven into a lei or other bodily adornment that can be worn by the dancers. As appropriate vessels, it is through the physical presence of the kino lau that the akua are entreated into the hālau.[10]

For hula practitioners like Pueo, the protocols for entering the forest, for moving within it, for speaking, for picking and collecting, are many and complex. At each step of the way, permission must be sought, and respectful rules of conduct observed. An intimate knowledge of the forest and other landscapes of the islands grounds this practice, a knowledge that is itself rooted in the traditional moʻolelo and mele that are taught and, in this way, passed down within the hālau hula. Far more then than a style of dance, hula is something more akin to "a way of life." As Kumu Hula Pualani Kanakaʻole Kanahele has succinctly put it: "This tradition teaches how to appreciate natural phenomena . . . love the land . . . and acknowledge and honor the presence of life. . . . Hula is a reflection of life."[11]

When I asked Pueo about his encounters with snails as part of this work, he explained that he had never actually seen kāhuli on Oʻahu. But on the island of Maui, where he was raised, smaller snails are sometimes still a common sight when they go to pick hala pepe (a plant) in the wet forest. He told me: "We check every single leaf to see if there are pūpū [snails] there. Every time we have to pull pūpū off and put them on a different leaf. We know that a snail might spend its entire life on that head of hala pepe so we have to replace it. . . . We're always watchful for them; we never want to hurt them."

Talking to Pueo, I gained a far greater appreciation of the religious and cultural significance of Hawaiian plants and animals. While Pueo's hālau emphasizes the unbroken continuity of their practice with that of the past, again and again in our conversations it was clear how profoundly the past two centuries of Euro-American settler and colonial presence have impacted not only on the specific practices of hula—which for periods was banned and forced underground to survive—but on the broader social, cultural, and environmental contexts of Kanaka Maoli life in these islands.

This chapter is an exploration of the historical and ongoing processes of colonization in Hawaiʻi. Of course, there are numerous ways in which this story might be, and has been, told.[12] In this chapter, unsurprisingly, I explore some of the ways in which the islands' snails have been caught up in these larger colonizing processes. My particular focus is the practice of shell collecting. While, as we will see, Kānaka Maoli have long collected the shells of Hawaiʻi's land snails, with the coming of Europeans and Americans this process was drastically transformed. Immediately, these new arrivals were entranced by these shells: from the explorers of the late eighteenth century, through the missionary period and the plantation era of the nineteenth, and on into those turbulent decades around the dawn of the twentieth century in which the Hawaiian monarchy was overthrown and a republic and then a US territory established in its place.

Throughout all these periods, unimaginable numbers of shells were gathered up and added to growing collections. The activity was enjoyed casually by everyone from school groups and scout troops to picnicking families and naturalist clubs. At the same time, there was a growing cohort of "confirmed shell collectors," mostly male and of Euro-American descent, who sought with more or less scientific training and rigor to assemble an inventory of the incredible gastropod diversity of these islands.[13] Accounts abound from the mid- to late-nineteenth century of naturalists and other shell enthusiasts traveling up into the forests of Oʻahu to collect snails, in some cases filling their saddlebags with shells.

As we will see, while it is hard to quantify the scale of the impact that shell collecting had on the snails themselves, it is highly likely that in particular times and places it was a significant one. As Mike Hadfield summed it up: "Those guys did a job, they wiped things out."

At the same time, however, shell collecting impacted on Kānaka Maoli. More precisely, it was one small part of a larger process of dispossession, alienation, and overwriting of traditional relationships with the ʻāina. The more I explored the history of shell collecting in Hawaiʻi, the more difficult it became to separate it from the larger story of European and American presence in these islands, one in which Hawaiʻi today remains a nation under US occupation, subject to the accompanying and ongoing social and cultural processes of settler colonialism.[14] This is a story rooted in more than 200 years of history, but one that brings us up to the present day. In this way, it is a story of the fracturing of relationships, but also one of ongoing resistance and survival. To borrow the words and sentiments of Kanaka scholar Leon Noʻeau Peralto: "This is not a moʻolelo of erasure," or at least not only that; it is one that seeks also to make present the "ways in which processes of erasure are questioned, resisted, and overcome."[15]

Many of the countless shells gathered up during this period can today be found in the Bishop Museum in Honolulu, as well as in a variety of other museums around the world. As will be discussed in more detail in the next chapter, these shells now play a vital role in the scientific research that is enabling us to better conserve those species that are left. In other words, they have become an important resource. But the history of their collection is nonetheless also a problematic one, a process that impacted in significant ways on both the snails and the native people of these islands. In this context, this chapter is an effort to keep these other stories in the frame.

I have started this chapter with a very different collecting story, one grounded in respectful appreciation of the forest as a realm of the gods, one in which snails are carefully avoided rather than being gathered up in bulk. This is not to say that snails were not impacted on by the activities of Kānaka Maoli; as discussed in the previous chapter, clearing for agriculture was probably significant in some places. Nor does it mean that snails have not sometimes themselves been collected by Hawaiians. But in holding these different practices and stories of collecting in tension, this chapter aims to cultivate a deeper appreciation of the profound and ongoing transformation in modes of understanding and relating to the ʻāina that has taken place on these islands in recent centuries.

BRAIDED LIVES

At the Bishop Museum in Honolulu, there is a remarkable object that offers us a portal into these entangled histories of shell collecting and colonization. Just a few days before my trip to Papahana Kuaola to meet with Pueo, I had been lucky enough to get the opportunity to go behind the scenes, into the museum's Ethnology Collection. Standing among rows and rows of cabinets in a windowless room, my curatorial guide, Marques Marzan, climbed a small stepladder and reached for a sturdy cardboard box about the size of a dinner plate on an upper shelf. Stepping down and walking over to a nearby bench, he gently placed the box down and lifted the lid. Inside, two interwoven strands of white and brown snail shells, patterned with intricate stripes and spirals, combined to form a stunning lei.

This is the one and only lei held in this collection that is comprised entirely of terrestrial Hawaiian snail shells, all of them of the genus *Achatinella*. Alongside its beauty, Marques explained that this lei is distinguished by its former owner: Queen Liliʻuokalani (1838–1917), the last monarch of the sovereign nation of Hawaiʻi. Little is known for certain about Liliʻuokalani's lei: how, why, or when she acquired it. What we do know, however, is that it was a somewhat unusual object. While marine shells were and still are very commonly used to make lei, the shells of land snails were far less frequently used in this way. It has been suggested that this may have been because the wearing of these shells was limited to females of high rank.[16]

In fact, kāhuli lei are so rare that most Hawaiians have probably never encountered one at all. When I asked Pueo about these shells, he told me a story about a friend who wore a land snail lei in a traditional hula competition. In such a context, he explained, things from the sea would not normally be combined with those that dwell in the mountains: "when you wear foliage as your lei ʻāʻī, your neck lei, you usually don't wear a seashell lei with it." His friend was marked down for wearing shells with a lei of maile—a forest plant—by one of the judges who assumed the kāhuli were seashells. But, Pueo explained, "mountain shells can grow on mountain foliage."

These kāhuli lei, even if rare, are a reminder that Kānaka Maoli have collected snail shells, perhaps for centuries, albeit probably on a modest scale. Alongside their ceremonial and decorative uses, occasional references indicate that some larger snails may also have been eaten by people, both raw and cooked inside the leaves of the ti plant.[17] But perhaps more importantly than any of these material uses, snails are potent hōʻailona (symbols or omens) in people's lives and stories, often indicating positive and righteous action and circumstance. This is, as we have seen, something that the stories tell us snails do most powerfully through their singing (chapter 2).

These relationships with snails are one small facet of the complex and intimate cosmology that has shaped Kanaka life on these islands for generations. This cosmology is rooted in a genealogical and familial relationship with the ʻāina. The plants, animals, and diverse creatures that reside in these islands are quite literally family for Hawaiians. This relationship is perhaps nowhere so eloquently expressed as it is in the *Kumulipo*, another part of the immense heritage that Queen Liliʻuokalani left to her people (initially printed in ʻōlelo Hawaiʻi by King Kalākaua, it was later translated into English by his sister and successor, the queen). This cosmogonic chant recounts the genealogy of this royal family, through generations of human and other than human life, all the way back to the slime mentioned in chapter 1: "The slime, this was the source of the earth."[18]

In this familial cosmology, humans are positioned as junior siblings, as "malihini (newcomers) in this universe."[19] In this relationship there is, as Kanaka scholar Jonathan Kay Kamakawiwoʻole Osorio explains, "a sense of real dependence in the way that children depend on parents and grandparents."[20] But it is, Osorio notes, also a relationship of obligation and responsibility, a kuleana. ʻĀina is "an older sibling, it cares for Kanaka by providing nourishment and is cared for in return."[21]

It is in this context that the previously mentioned kino lau (god forms) must be understood. Far from being removed from the world of living beings, Hawaiian akua (deities) are an integral part of this familial landscape: "humans are part of a vast family that includes celestial bodies, plants, animals, landforms, and deities."[22] As Pueo summed it up: "In the forest we are surrounded by our ancestors. We are descended from these akua so these are my relations."

Through innumerable generations since their arrival, Kānaka Maoli worked to cultivate relationships of reciprocal care and nourishment with the plants and animals, the waters and lands, the ancestors and the akua, of the Hawaiian Islands. They did this in their intricate systems of kalo (taro) and ʻuala (sweet potato) cultivation for diverse environments, including irrigation systems where necessary; in their massive walled fishponds; in the construction of their waʻa (canoes), homes, and heiau (temples). In living on, with, and for the ʻāina, they developed intimate systems of knowledge and attention related to their extended, more-than-human familial networks.

Snails, of course, are a part of this family, and Hawaiian moʻolelo and mele are thick with snail knowledge. Snails go by a variety of Hawaiian names in these sources. To an outsider like myself, at least initially, these names swirled in my head and were no easier to grab hold of than the long Latin-esque scientific ones that these snails have collected over the past couple of centuries. We have already encountered the two most common Hawaiian names for snails: the first is kāhuli, a term that also means to turn or change and is generally thought to refer to the way in which a shell rocks from side to side as its snail travels along; and the second is pūpū-kani-oe, literally meaning "shell sounding long." While the name kāhuli is reserved for land snails, and by some people just for the larger, more colorful ones, the term pūpū (without the distinguishing kani-oe) is used to refer to snails, or indeed shells, including marine shells, of any kind.

But many other names also exist. Aimee You Sato and her coauthors have conducted a review of historical Hawaiian language sources and carried out interviews with cultural practitioners in a search for these names. They tell us that snails were at times also called pololei, which translates as perfect or correct, and hinihini, meaning "delicate voice." These names can also be modified to provide more specific refences, as in the case of pūpū-moe-one, "shell that sleeps in the sand"; pūpū-kuahiwi, "mountain shell"; and pololei-kani-kua-mauna, which refers to perfect singing in the mountain ridges.[23]

Countless other modifications of these names probably exist, such as hinihini konouli, which refers to a shell of a dark shade, and hinihini kua mauna, the voice of the mountain ridges.[24]

In my discussions and research, I have not been able to locate a larger Kanaka Maoli system of classification for the islands' diverse snails. It seems likely, however, that such a system would have existed, as it did for many plants and birds. Noah J. Gomes has discussed some of the key historical examples of Hawaiian classification systems for birds.[25] While these systems differed somewhat, they seem to have primarily divided birds into groups on the basis of their utility to people, the environments in which they were found, and "the function that the bird has to the greater ao holoʻokoʻa (the world), that is to say, the function the bird has in the ecosystem."[26]

The names given to particular bird species also tended to follow a particular pattern. Specifically, birds tend to have Hawaiian names that reference (1) a peculiarity of their appearance, (2) a sound they make, (3) a distinctive behavior or habit, or (4) a historic or legendary person who is in some way associated with the species.[27] Clearly, many of these naming conventions also applied to snails. If these names are anything to go by, which they surely must be, then Kānaka Maoli have long paid considerable attention to the shells, movements, habitats, and much more of the many snails that make their homes in these islands.

FROM A LEI TO LAND SHELL FEVER

The period of modest collection and use of snails by Kānaka Maoli ended abruptly with the arrival of Europeans and Americans.[28] Another lei, this one also comprised entirely of shells of the genus *Achatinella*, marked the beginning of this period. This lei was purchased by the English Captain George Dixon on Oʻahu and taken by him back to London in 1786. The shells that comprised the lei immediately attracted the attention of collectors. The lei was cut up and sold for parts, with each shell reportedly fetching the princely sum of $30 to $40.[29] The fate of these individual shells is, for the most part, unknown; they did, however, have a lasting legacy, bringing Hawaiʻi's living jewels to the attention of the wider world for the first time.

This was a turbulent period in Hawaiian history. Dixon's arrival took place a mere eight years after Captain James Cook's 1778 visit had put these islands onto the maps of Europe (in fact, Dixon had also been on that earlier voyage). As was often the case with European explorers during this period, Cook determined that he might provide a more appropriate name for these islands and so they did not appear on these maps under any of their local names, but as the Sandwich Islands, named for Cook's patron the Earl of Sandwich, John Montagu. This was a time of intense warring between various mōʻī (paramount chiefs) and aliʻi of the islands, each vying for control of territory. In the 1780s and 1790s, Kamehameha (1736–1819) rose to prominence to unify all of the islands under his rule. The Kingdom of Hawaiʻi was born, taking the name of its largest island and Kamehameha's birthplace.

As discussed in the previous chapter, the arrival of outsiders during these decades transformed the Hawaiian landscape initially through the introduction of new species like cattle and goats, as well as through the breakdown of agricultural and irrigation systems that resulted from the incredible loss of human life to new diseases. Throughout this period of change, the islands' snails continued to capture the attention of Western visitors. Naturalists, collectors, and other curious people of all kinds purchased or gathered shells to be taken back to Europe or America, where they were often snapped up by museums or private collectors.

Then, in the 1820s, life in Hawaiʻi changed again. Christian missionaries from Boston began to arrive and settle in the islands with their families. The influence of Christianity contributed significantly to the undermining of the kapu system—the traditional rules that ordered Hawaiian society—which had already been weakened by what Osorio has referred to as "the great dying" of the preceding decades.[30] These new arrivals increasingly took on important political, economic, and cultural roles.

For the islands' snails, the presence of permanent American settlers also introduced a dramatic new condition. People of all ages took up shell collecting. Children and adolescents of missionary families in particular were encouraged to do so. The activity was viewed as a healthy outdoor pastime, but, importantly, one that was also an "improving kind of work, strenuous and educational."[31] For these predominantly New England Christians, a close interest in the natural world was also a kind of religious activity: attention to God's creations provided a tangible means of better understanding the divine.[32] John Thomas Gulick, the missionary and naturalist whose theories about nonadaptive evolution we encountered in the previous chapter, was one such person. In his words, shell collecting enabled "the study of God's character as displayed in his stupendous works."[33]

In writings of this period there is frequently a close alignment between, on the one hand, natural history collecting and the effort to name and catalogue the world and, on the other, the betterment of a people and their society. We see this strikingly in an 1837 lecture by the president of the newly formed Sandwich Island Institute. The lecture opens by outlining the objectives of the institute, centering on the "intellectual and moral improvement of its members," and concludes with "an earnest solicitation to every member to endeavor to make collections of every substance interesting in natural history."[34]

By the 1850s, shell collecting in Hawai'i had gotten more than a little out of hand. In his recollections of the period, David D. Baldwin (1831–1912) noted that at the time "an intense interest was awakened on the islands." Locals started referring to it as "land shell fever."[35] The fever, it seems, especially afflicted younger males. Baldwin, in his early 20s, was himself caught up in it and seems never to have recovered: while he went on to become a successful politician and businessman, during his lifetime he was also widely recognized as one of the foremost authorities on Hawaiian land shells. When his daughter married in 1885, the newspaper reported that her wedding reception at the family home was arranged so that guests might view his "large and magnificent collection of shells."[36] There is no discussion of the happy couple's views on this arrangement.

In her account of Hawaiian natural history and the missionary families during this period, biologist E. Alison Kay has noted that "by the early 1850s virtually all the boys had succumbed" to "conchological fever," and as a result, as the Reverend Alexander put it in 1852, they "daily traverse the ravines in quest of land shells."[37] Numerous reports exist from this period of people traveling into the mountains for a few hours or even days of snail-centered fun, returning with huge quantities of shells. The Punahou School in Honolulu, established about a decade earlier to educate the children of missionaries, seems to have been an important hub for this activity. Articles in the school newspaper in March 1853 tell of a picnic in which over 4,000 shells were collected, and an expedition into the mountains in which roughly half that number were again gathered up.[38] Today, there are fewer individual snails of a whole range of species left on the entire island of Oʻahu than were collected by some of these groups over a period of a few hours.

Perhaps the most well-known of those afflicted by land shell fever during this period, though, was Gulick. As a young man he amassed an incredible collection of *Achatinella* shells. The vast majority of his collecting was done in the 3-year period from 1851 to 1853, just before he set off for college on the US East Coast. When he left, he took many of his shells with him. Indeed, he continued to cart a selected reference collection from them around with him over the following decades to his various missionary posts in Asia.[39] It was from these collected shells that his evolutionary thinking developed.

Entries in Gulick's journal from 1853 provide us with an interesting insight into his collecting process. In particular, they make clear that a great deal of the actual labor was undertaken by others, specifically Kānaka Maoli living in villages around the island. In some cases, Gulick paid local people to collect with him. For example, on July 27, 1853, he reports that in the company of two friends "and eight natives [he] took an excursion to the woods in quest of shells." In a single valley they managed to collect over 1,400.[40] In other cases he enrolled local children as helpers. But many of the shells in Gulick's vast collection were acquired through what seems to have been a significant network of Kanaka Maoli villagers collecting on their own. Gulick, riding between villages, slowly filled his saddle bags with shells as he went, leaving information about when he would next return.

Gulick's diary also makes clear that as he collected shells in the summer and fall of 1853, Kānaka Maoli were dying all around him in large numbers, afflicted by an actual fever, smallpox. Thought to have been introduced on Oʻahu in that year, the disease quickly spread, infecting thousands of people across that island, as well as on Kauaʻi, Maui, and Hawaiʻi. As many as 6,000 people are thought to have died within a single year.[41] According to historian Seth Archer, "Honolulu became a charnel house. Government wagons carted the sick and dead through town, and yellow flags hung in doorways."[42] Gulick was not indifferent to the plight of the sick and dying: he delivered medicine as he traveled around collecting shells, occasionally noting that the people he encountered were in his prayers. But his resolute focus seems to have remained on the snails.

In September 1853, Gulick visited the area of Moanalua and found the local people "scattered like sheep" after several were taken by smallpox. Leaving some medicine for them, he headed on. In his diary he recorded: "Leaving Moanalua I passed on through Ewa stopping at different points, . . . and finally passed up the valley of Waiawa, but found no shells ready for me in any of these places, for the natives have been sick and dying, and the living busy burying their dead."[43] After a restless night of dreams "filled with piteous groans and implorings of the sick," the following morning he was at it again: "myself and two of the native men were in the woods from eight o'clock till late in the afternoon; and when we returned we had from 200 to 400 each."[44] The smallpox epidemic finally came to an end in January 1854. By this time, Gulick was at sea on his way to college.

For many Kānaka Maoli these years around the midpoint of the century also marked the culmination of a long period of dispossession. From 1848, a process known as the Māhele transformed the traditional and customary rights in land that had structured Hawaiian society in order to create a more conventional Western system of private and public property. This change was driven by a variety of factors, including the threat of invasion, massive declines in population that had left much of the land untended, and growing pressure from settlers and aliʻi who wanted to be able to buy and sell land as part of their commercial endeavors.[45]

Many of the common people of Hawaiʻi, the makaʻainana, strongly opposed this change. Kanaka scholar Davianna Pomaikaʻi McGregor has argued that "throughout history, the Hawaiian people maintained a deep abiding faith in the land as the source of physical sustenance, spiritual strength, and economic well-being." It was with this faith in the ʻaina that, McGregor notes, the people petitioned the king, Kamehameha III, not to sell land to foreigners, writing: "The land strives (kūlia) for revenue every day. The earth continues to receive its wealth and its distinction every day. There would be no end of worldly goods to the very end of this race. But, the money from the sale of land is quickly ended, by ten years time."[46]

But the changes came, nonetheless. Although maka'āinana were given an opportunity to claim small parcels of land, it did not work out at all in their favor. Initially, most private lands were owned by the king and a small group of ali'i. Gradually, however, more and more land was acquired by wealthy outsiders, especially for the establishment of large ranches and sugarcane and other plantations.[47] A process had begun through which, as the Kanaka anthropologist J. Kēhaulani Kauanui has succinctly put it: "Hawaiians and their descendants became largely a landless people."[48]

When I asked Sam 'Ohu Gon about this process, he told me that it was important to remember that this period of mass mortality and dislocation from the 'āina was also a period of profound "cultural disintegration." Sam is a senior scientist and cultural advisor for the Nature Conservancy Hawai'i, and a distinguished Kumu oli, a singer and teacher of chants. He noted that in a culture in which knowledge was transmitted across the generations in oral form, the death of so many people, so quickly, had a profound impact. Kānaka Maoli were painfully aware of this loss at the time. When the Christian missionaries introduced a written alphabet and the printing press, Hawaiians quickly took to these new developments. In the latter half of the nineteenth century, Hawai'i was one of the most literate nations on earth.[49] Kānaka Maoli turned these new tools to their own purposes, establishing a series of Hawaiian-language newspapers that, alongside the news of the day (including opposition to interference in the nation's political system), became repositories for traditional stories and knowledge, helping to ensure that they would not be lost.[50]

Bound up with these profound impacts on Kānaka Maoli were widespread transformations of the 'āina itself. As Sam put it: "When you see the decline in population and knowledge, that same kind of decline is being experienced by the native plants and animals of Hawai'i, particularly in the lowlands where agriculture and ranching and introduced predators like rats and dogs essentially replaced entire ecosystems." For a people whose lives and familial relations were so intimately tied to the 'āina, these changes mattered profoundly: "When your universe is created out of an assemblage of interacting physical manifestations of gods that essentially provide for all your needs as long as you provide for their care as well, to see that disintegrate around you is a profoundly sad experience."

As the nineteenth century drew to a close, control and ownership of further vast areas of land, and then of the nation itself, was taken from Kānaka Maoli. The last monarch of the sovereign nation of Hawai'i, Queen Lili'uokalani, was overthrown in 1893. The queen and her government were ousted by a group of mostly American citizens and Hawaiian subjects of American descent, with the assistance of US Marines from the USS *Boston*, which was docked in Honolulu at the time. The hope of the instigators of this overthrow had been that the United States would annex the islands, but concerted opposition from the queen and the people prevented this.[51] Instead, a republic was formed, headed up by a white oligarchy. But it was short lived. In 1898, with a change of government in Washington, DC, the Hawaiian Islands were annexed to the United States as part of its growing Pacific empire.

NATURAL HISTORY COLLECTING AND THE SCIENCES OF SETTLING

Throughout these decades of turbulence, the gathering of snails from the islands' forests continued at pace. Newspaper articles from the period tell of school groups and clubs who actively pursued the pastime; others advertised shells for sale, announced land shell exhibitions and competitions, or reported on collecting adventures. In 1911, an article in the *Hawaiian Star* goes so far as to insist that "Boy Scoutdom should have some more useful function in Hawaii than the mere collecting of land shells."[52] In diverse ways, more and less obvious in these accounts of snail-related activities, we see how shell collecting became part of the larger domesticating processes that "settled" new arrivals and then occupiers and colonizers into these islands.

Decades of historical scholarship have shown that all over the world during this period, imperial, colonial, and nationalistic endeavors were caught up with the work of natural history: of collecting, cataloguing, studying, and displaying the biological and geological richness of the earth.[53]

Sometimes specimens were sent back to distant centers of scientific power like London, Paris, and Washington to bolster knowledge and provide useful new resources. Other times, however, they played a more local role, contributing to what the historian of science Libby Robin has called the "sciences of settling," which translated and transformed local places to suit the expectations and needs of newly arrived peoples.[54] From crop development to local watershed management, the natural sciences contributed to the transformation of Hawai'i during this period.[55] But there is a largely unexamined history to tell about the way in which shell collecting—alongside bird watching, hiking, and other "nature activities"—played a subtle but important role in shaping the cultural and political context of Hawai'i during this period.

One particularly striking example of this settling work took place in 1895. Two years after the overthrow of Queen Lili'uokalani and mere months after the establishment of the short-lived republic, the newly established Bishop Museum announced an "Exhibition of Land Shells." To be held later in the year, the event was an open competition for residents of the islands to display their shell collections. Announcing the exhibition in the *Pacific Commercial Advertiser*, the first director of the museum, William Tufts Brigham (1841–1926), noted:

> There are many young persons who have made very creditable collections, and it is hoped that a more thorough enjoyment of this interesting pursuit may come from a careful and accurate arrangement of the spoils of many a mountain excursion. By comparison of collections each one may be stimulated to enlarge his own.[56]

The trustees of the museum threw their support behind the event, encouraging local people to make public "the results of their researches and labors." They concluded their short statement in the newspaper by noting that some collectors in the islands "have had the honor of having their names attached to new species, which they have discovered."

Names for snail species, it seems, played a particularly important role in this planned exhibition, connecting the work of collecting to scientific knowledge production and distinguishing it from the mere acquisition of trinkets. In his instructions to entrants, for example, Brigham noted: "shells may be arranged after the owner's individual taste, but specimens must be named and localities given." Of course, the names to be provided here were the newly acquired scientific ones, not Hawaiian names—in fact, the following year the government of the republic banned the teaching of 'ōlelo Hawai'i in schools.

Naming is always a powerful, formative work. As we will see in the next chapter, the effort to identify and name snail species is a vital part of contemporary conservation efforts. But naming is more complex than this. Among other things, it is frequently also an act of dominion, one that more or less intentionally asserts the authority to categorize and order reality.[57] When taxonomy takes place within the complex folds of an ongoing process of occupation and colonization, it becomes all the more fraught, as new names and understandings are layered over the old, perhaps supplanting them or even coexisting, but inevitably changing the way in which people understand and relate to the world around them.

Kānaka Maoli at this time were living through waves of mass sickness and death, through the overthrow of the monarchy and severe disruption of the social order, including the large-scale transformation of island environments and their more-than-human communities of life. Osorio has described this process as a "dismembering of lāhui," a dismembering of the people and the nation in their close association with the lands and waters of the islands. Language is an intimate part of this process of dismemberment, as 'ōlelo Hawai'i was displaced and other ways of speaking and naming the world took its place, overwriting the landscape and people's relationships with it.

This is a process that began in some ways right at the outset of European contact, with Cook's designation of the Sandwich Islands, a name that initially stuck and spread but was eventually supplanted in the mid-nineteenth century.[58] But many other newly given names from this period are still with us. In fact, during this time the young sons of missionaries "were wont to ramble high in the hills around Mānoa [Valley], naming the various peaks" as they collected shells and ferns.[59] Names from this period include Round Top, Sugar Loaf, Olympus, and the area where I went walking as described in chapter 2, once called Pu'u 'Ōhi'a but now known as Mount Tantalus by most residents of Honolulu.[60]

There is an unavoidable overwriting of the landscape at work here, a reordering of reality. Names matter. As Kanaka poet and scholar Brandy Nālani McDougall has put it: "We have always known that words have immense power. The 'ōlelo no'eau [proverb] 'I ka 'ōlelo nō ke ola, i ka 'ōlelo nō ka make' (In words there is life, in words there is death) is just one of several examples that recognize how words have the power to actualize, to be either life giving or destructive."[61] Place names, in particular, are important in Hawaiian culture; as Kanaka scholar ku'ualoha ho'omanawanui notes, they carry connections and identity, linking people to their home, their family, and their (hi)stories.[62] In this way, to alter them is to unsettle these relationships.[63]

But smaller features of the landscape, including snails, also received new names at this time. Many of these names were those of the American families who had been instrumental in the transformation of this place. *Achatinella dolei* is a case in point. Named by Baldwin in 1895, he noted at the time: "We take pleasure in dedicating this beautiful shell to His Excellency S. B. Dole, First President of the Republic of Hawaii."[64] But this is one species among many. Baldwin, Byron, Judd, Cooke, Spalding: the list of species names within the genus *Achatinella* reads like a who's who of wealthy and powerful settler-colonial families, many of them with their roots in missionary days but who went on to senior positions in politics, commerce, and law.[65]

Other snail names more or less subtly redid traditional Hawaiian notions. Take, for example, *Achatinella vespertina*, a naming that explicitly references stories of singing snails, but that converts their song into the Christianized form of a vesper.[66] These new snail names must have played some role—perhaps only ever a small one limited by the reach of taxonomic nomenclature—in further alienating Kānaka Maoli from their lands and traditions.

Of course, at the same time as they disconnected Kanaka ways of knowing, these acts of naming also rendered the land familiar and at least potentially knowable and relatable in new ways to settlers. We catch a glimpse of this process in an entry in the diary of the young missionary daughter Lucy Thurston in 1835, when she excitedly reports on an outing in which "we gathered nearly a quart of Stewart shells . . . [named *Achatinella stewartii*] as Mr Stewart was the first who ever carried them to America."

For many people of the time, these colonizing dynamics were surely not at the forefront of their interest in natural history. But for others, they clearly were. For some people natural history, perhaps especially collecting and display, were part of a deliberate effort to transform the nation. In the *Pacific Commercial Advertiser*, just a few days after the announcement of the land shells exhibition in 1895, the editor of the newspaper, Wallace R. Farrington, endorsed the event as a "preliminary move" in the generation of interest in the distinctive natural features of the islands. This he insisted, was a key part of building the nation, at this stage an insurgent republic:

> An interest in Hawaii as Hawaii, its political history and its natural history, its particular personality in all things, is a feature which ought to be given a growing importance as the Hawaiian-born Anglo-Saxon, Latin and Asiatic becomes more a factor in directing the affairs of the nation.[67]

Farrington moves seamlessly from this discussion into an argument for the "supremacy of an English-speaking people in the islands," a supremacy grounded in his view that "Order is the basis on which every English-speaking society rests. It is the primary instinct of the English-speaking race." While he is here explicitly speaking about the rule of law, about a political order—a highly ironic statement in the context of the recent overthrow of the monarchy—there is a clear connection to his discussion of natural history. What better demonstration of this capacity to order the world, or at least these islands, than a neatly arranged, carefully labeled cabinet of shells?

"SHELLING IS NOT WHAT IT USED TO BE"

In the first half of the twentieth century, the shell-collecting craze gradually subsided and then petered out. During a period of a little over 100 years, though, this practice must have had a significant impact on the islands' snails. Ultimately, we will never know just how many snails were pulled from Hawai'i's forests during this period. Looking back at the limited (English-language) historical records that exist, Mike Hadfield has noted that they "lend visions of tens of thousands of gastropods being frequently reduced to 'shells.'"[68] Indeed, many private collections were in excess of 10,000; William H. Meinecke is said to have had more than 116,000.[69]

It is similarly difficult to determine how significant the impact of all this collecting was on these species. In at least some cases, for some species, it seems highly likely that collecting was a crucial factor in their decline. Species like those of the genus *Achatinella*, with their slow rates of reproduction, limited mobility, and highly localized distribution, would have been exceedingly vulnerable. As Mike put it: "Predators as selective as human shell collectors could . . . drastically affect such isolated populations with ease."[70]

Despite this fact, throughout this whole period of intensive collecting very few collectors seem to have worried much about their own impacts, even as they noted repeatedly that some snail species were becoming rare and had perhaps even disappeared altogether. Instead, the presence of grazing cattle and predation by rats were generally seen as being to blame for this decline.[71] It's likely that these introduced animals, alongside habitat loss through urban development, ranching, plantations, and more, were the principal drivers of loss during this period. And yet, in particular pockets of place and time, for particular species, shell collecting may have mattered profoundly.

Among the accounts of hobbyist collectors, I have found no indication that the increasing rarity of snails encouraged them to cease or modify their practices. Newspaper articles from the period effortlessly combine a recognition of snail scarcity with the hope that young people's interest in collecting might be further stimulated. In an article in the *Hawaiian Gazette* in 1873, reporting on a donation of shells by Gulick to Punahou School, his alma mater, the school takes the opportunity to say that "we hope there will be awakened a new interest in these shells from the mountains, many of which are now most or quite extinct."[72] As we have seen, a quarter of a century later, similar ambitions were tied to the Bishop Museum's shell exhibition.

While this lack of concern about extinction seems highly unusual from a contemporary perspective, it was in no way restricted to snails in Hawai'i at this point in history. As the historian of science David Sepkoski has argued, while popular and scientific understandings in Europe and America shifted considerably during the latter half of the nineteenth century and into the twentieth, extinction was by and large seen as an unavoidable inevitability as species outcompeted one another or succumbed to their own inbuilt "racial senility" (with similar ideas frequently applied to non-Western peoples). As a result, extinction was rarely seen as a problem to be solved or a situation that should, or could, be intervened in.[73]

And so, even in the case of snails that are, for all they know, the last individuals of their kind, collectors often went on with their work. For example, in a report on an outing of the Gulick Natural History Club in 1919, we are told: "Shelling is not what it used to be, for shells are very scarce in this section of the Koolau Range. We managed to get seven specimens of *Achatinella viridans*: we considered this a lucky find." A little later the party comes across 12 snails of the species *Achatinella phaeozona*. The search seems to have been exhaustive, with some found "near the tops of the kukui trees." This find, we are told, "was a surprise to us, for we were told that these shells were extinct in Keawaawa." They too seem to have been gathered up and taken back to the Bishop Museum for identification.[74]

The impacts that these collectors had would have depended on their collecting practices. Sadly, this is something that we also know very little about. Were most collectors just taking the shells they could reach from the ground or climbing trees? Were they taking only the nicest shells, or everything they could find? Were they just taking adults or snails of all ages? In his instructions to young people about this "hobby" in 1900, Baldwin describes using a hook on a pole about 2.5 meters long to pull down taller branches to access their snails. At the same time, he also says: "I beg young friends never to collect young or half-grown shells. They are difficult to clean nicely, and are of little or no value as specimens. Leave them on the trees to grow for the benefit of future collectors."[75] It's hard to know whether we should see this plea as a positive sign, that there existed a conservation ethic around disappearing snails, or rather focus on the need that drove this plea and perhaps see it as an indication that collectors were cleaning out parts of the forest. Either way, however, his injunction to preserve was justified not with reference to the good of the snails or the environment, but rather for the benefit of future collectors.

One would hope that among the more scientifically inclined, greater efforts were made to ensure that their collecting was not detrimental to snail species. This certainly doesn't seem to have always been the case, though. Gulick's own characterization of his efforts as "ransacking" the forest doesn't fill one with confidence. Reading accounts by leading scientists writing in the early twentieth century, I have found myself, on more than one occasion, in a kind of shocked rage, yelling at the page, "Then stop collecting!"

In the *Manual of Conchology* edition from 1912–1914, Henry Pilsbry, head of the Department of Shells at the Academy of Natural Sciences in Philadelphia, reports "a fact which I find is known to all Island collectors, that a species may be found on one tree or shrub, year after year, without spreading to neighboring shrubs." He then goes on to say:

> It strikes the outsider as uncanny to be taken to one certain tree to collect specimens of some snail which either has never been found anywhere else, or nowhere else in the neighborhood or district. To find that trees just as good all around are barren of specimens is always a surprise.

While this extremely limited distribution was a surprise to some, it seems not to have been one that was able to mitigate against taking collectors to these trees in the first place. Even understanding these patterns of distribution, many collectors seem not to have worried significantly about their own activities. Or, if they did, they viewed the work that they were doing as important enough to override such concerns. Repeating a dynamic that has frequently been observed of cataloguing and collecting projects of all kinds, the knowledge that snails were in decline, instead of staying the collector's hand, often seems to have functioned as a call to action, to ensure that a valuable scientific specimen was catalogued before it was lost forever.[76]

SNAILS, 'ĀINA, AND A CULTURAL RENAISSANCE

Pueo and I walked down a muddy path toward a gently flowing stream. Having finished our interview, he had offered to show me the lo'i, the kalo (taro) fields. Still on the property of Papahana Kuaola, the fields were just a short walk away. We crossed over the stream, and as we came up on the other side, spread out in front of us were several large terraces, each one composed of a series of long mounds of soil supporting a row of kalo plants.

The particular approach to cultivation taken here, Pueo explained, is not the one you might take on the other side of the island, or even in the next valley. Rather, like all Hawaiian knowledge, the approach is a thoroughly place-specific one. Much of this knowledge is encoded in traditional stories. In this case, Pueo told me, the stories inform us that kalo should be cultivated on mounds in this valley, a place, he noted, in which the water is spring fed and therefore higher in minerals but lower in oxygen, and so plants benefit from the additional elevation of their roots.

These loʻi are at one and the same time an educational space and a working farm. Here, students learn how to observe and take care of the ʻāina; they learn how to cultivate kalo in a way that draws on diverse knowledge systems, from traditional Hawaiian practices and stories to the sciences of hydrology and agronomy. At its core, Pueo told me, this educational program is an effort "to cultivate these students to be the problem solvers for Hawaiʻi in the next generation." For perhaps 1,000 years, Kānaka Maoli have lived with and managed these islands. The insights of Hawaiian approaches, he explained, must be a central part of grappling with the problems we now face, from food and water security to biodiversity loss and rising seas.

The work undertaken at Papahana Kuaola is one small part of a broader renaissance of Hawaiian culture that has been taking place in the islands over the past several decades. There are many strands to this process, from the recovery of navigation and voyaging knowledges and practices to the revitalization of Hawaiian music, hula, agriculture, and more.

ʻŌlelo Hawaiʻi, the Hawaiian language, has also been a key part of this renaissance. For about four generations following the ban in 1896, it was not taught or spoken in schools. Over the past several decades, however, a host of dedicated people have worked to ensure that the language has again become a vibrant and dynamic cultural force; today, students can receive an education in ʻōlelo Hawaiʻi from preschool to university. In Hawaiʻi, we see that our current time of extinctions is not only that. It is also, in important ways, a time of resurgence, of cultural renaissance, for the Hawaiian people.

ʻĀina has been inseparable from this renaissance, as it is from culture itself. Struggles over land—efforts by Kanaka communities and their allies to retain or regain ownership or access—were, as Kanaka scholar Haunani-Kay

Trask has noted, key to "the birth of the modern Hawaiian movement."[77] This has been the case across diverse contexts, ranging from rural communities in places like Kalama Valley on Oʻahu to military training and test sites like the island of Kahoʻolawe (chapter 5). At the same time, for many Kānaka Maoli today, protecting the ʻāina—including its community of plants and animals—is seen as vital to the continuity of culture; while cultural practices—from traditional stories to methods of kalo cultivation and the correct protocols for hula collecting—are themselves viewed as being at the heart of a distinctive Hawaiian form of sustainability and conservation that is increasingly being integrated into both education and natural resource management across the islands.[78]

Grounded in the ongoing reality of colonization, many Kanaka scholars and activists understand a central strand of this work to be a process of "reconnection" to the ʻāina and to Hawaiian culture after more than a century of suppression and dispossession. Pueo summed up the situation simply as we stood by the loʻi. In his view, it is vital to thicken these connections, and this is a key part of the work he does with his students:

> If they don't have a sense of pride in, or responsibility to, this place, and a good foundation in Hawaiian identity, there's not that much that is keeping them here. About eight residents of Hawaiʻi leave every day to move to another place because it is so expensive here, and to be honest, there's not really a connection for many people to place anymore, unless they were raised in a traditional family.

Kanaka scholar Kali Fermantez makes a similar point when he notes: "In Hawaiʻi today, Native Hawaiians find ourselves literally and figuratively out of place. Such displacement can be countered through conscious acts of re-placement, or reconnecting to Hawaiian ways of knowing and being that are rooted in place."[79]

This work of (re)connection draws on a variety of deep cultural sources. While much was lost or damaged by processes of colonization, much also endured. In a similar vein to Indigenous thinkers in other parts of the world who have emphasized "survivance,"[80] Kanaka scholars like Davianna Pomaikaʻi McGregor have worked to identify "cultural kīpuka" that were sustained into and through the darkest years. The Hawaiian term "kīpuka" refers to the small sections of forest that are often left standing after a lava flow moves through a landscape. It is from these preserved areas that seeds, spores, and other forms of life are able to spread out, to reforest the barren landscape. Building on this understanding, McGregor has argued that key rural sites, often isolated and overlooked, enabled Hawaiian language, culture, and relationships with the ʻāina and akua to continue well into the twentieth century.[81]

Alongside these particular geographical locations, other scholars have identified specific traditions, like hula, that were better able to carry knowledge forward, even if in an altered form or underground, as well as specific material resources, like the vast body of newspaper and other archival materials in ʻōlelo Hawaiʻi.[82] In their own ways, all of these diverse kīpuka have safeguarded and preserved invaluable cultural resources for subsequent generations of Kānaka Maoli.

Snails play their own small role in this bigger story of colonization and Kanaka (re)connections to the ʻāina. When I asked Sam ʻOhu Gon about the significance of snail extinctions for contemporary Kānaka Maoli, he explained that they are but one example of the profound challenges and possibilities of this time. "Nowadays," he told me, "when a plant or animal is declining, there is a recollection of past declines." With so many species now gone or exceedingly rare—and not just snails but plants, birds, and a host of others—it becomes that much harder, but also that much more important, to enhance connections to culture and ʻāina. In the not too distant past, all of the residents of Oʻahu would have been able to encounter the large *Achatinella* tree snails; they may well have had clusters of them in their yards, he said. At that time, "If you sang about snails, people would have been intimately familiar with them." Now, however, many people know little to nothing about them—if they have seen these snails at all, it is probably only in a photo—and at the same time, "they may not know the traditions and significance that they once had."

That significance, which snails still hold today in different ways for many—even if not all—Kānaka Maoli, is grounded in Hawaiian cosmology.

In this context, the extinction of snails, alongside the disappearance of other plants and animals, is the loss of family members, of siblings to which living generations have a responsibility. It is also, the loss or diminishment of a connection to the akua who are made manifest in the world in these creatures. As snails and other plants and animals disappear, Pueo explained, they take part of the religion with them. "If we no longer have access to the bodyforms of our ancestors, of our akua, then that form of worship stops."

At the same time, snails and other creatures anchor Hawaiian mo'olelo and cultural knowledge in the world. These stories tell of the history of the islands; they give guidance in the conduct of life, insight into the labors of various professions, and much more. One particularly well-known snail mele (chant), which has even been turned into a popular children's song, offers a beautiful example of the richness of these accounts. The mele, usually referred to as "Kāhuli aku," is often translated as a description of a kāhuli (snail) and a kolea (a bird, the Pacific golden plover), both of which are interacting in different ways with the 'ākōlea fern. In this way it encodes ecological knowledge.

But all of these terms have diverse meanings in 'ōlelo Hawai'i, so there is more going on here. When I asked Pueo about this chant, he pointed to another possible layer of meaning (kaona), saying that in his tradition it is a mele that encodes an act of sexual intimacy through the metaphor of collecting water. From this perspective, the term kolekolea is not (only) a reference to the singing of a bird and/or a snail but is a description of young people teasing and chasing one another in the forest. The mele alludes specifically to adolescent love, the kind that might readily have been struck up around common water sources shared by various families. As Pueo explained: "Young kids were sent to fetch the water, adults don't fetch water."

In this way, stories and chants like this one—perhaps about snails, perhaps (also) about other things entirely—become ongoing sources of inspiration and reflection on the entwined ecological and social relationships and histories of these islands. While some of this insight is held onto in the mo'olelo and mele themselves, these are systems of understanding that retain their meaning in dialogue with a world of creatures and relationships. As those relationships break down, the stories must inevitably, eventually, begin to lose layers of meaning.

The profound entanglements between Kānaka Maoli culture and the 'āina mean that the loss of plants and animals is inherently also a cultural loss, or at the very least a wound. The loss of snails—like that of so many other species—is the loss of ancestors, of religious practices, of connections to gods, of cultural and ecological knowledge, and much more. At the same time, however, this situation means that the revitalization of culture and of the 'āina can also often nourish one another. In the context of the islands' gastropods, we are seeing that cultural knowledge of snails is tangled up with their fleshy living forms, both holding each other in the world, vulnerable and at stake in one another.

Right across the islands, Kanaka scholars, artists, activists, hula practitioners, and their allies are working, dancing, singing, and thinking with snails. They are drawing on traditional snail stories, and they are weaving their knowledge into conversation with contemporary stories and realities. In chapter 5, we will encounter one particularly powerful form of solidarity between snails and Kānaka Maoli, in the form of struggles to protect and restore Mākua Valley.

The work of collecting shells and naming snail species that has been the key focus of this chapter also has a role to play in this work of (re)connecting to the 'āina. When I asked Larry Lindsey Kimura about Hawai'i's snails, he told me that the first time he encountered them was as a high school student in the early 1960s in the form of empty shells at the Bishop Museum, shells that may very well have been collected by missionary sons. Larry is a professor at the University of Hawai'i at Hilo, and is sometimes referred to as the grandfather of Hawaiian-language revitalization. For over four decades he has worked to preserve and record 'ōlelo Hawai'i, while also ensuring that it remains a dynamic, living, spoken language among new generations. As part of this effort, Larry helped to establish the Lexicon Committee, a group of language and culture experts whose role it is to develop new Hawaiian names for everything from telescopes and computers to children's toys. As a result, Larry is a person who has spent more time than most thinking about the importance of naming.

Most of Hawai'i's snail species, like other invertebrates, have not been given Hawaiian names. Partly this is just because there are so many of them, but also, Larry explained, it is a question of prioritizing the terms that people use in their day-to-day life: for the vast majority of us, 750 snail names would be excessive, a fact also evidenced by the lack of English common names for almost all of these species. In addition, however, there is no reason that Hawaiian names ought to mirror contemporary taxonomic divisions, which are themselves changing. The kinds of distinctions, the ways of dividing up the world, that work for scientists aren't necessarily those of other cultures.

When I asked Larry about the scientific names given to snails both in the past and today, he noted that there is considerable room for improvement in this practice. Names such as *Achatinella dolei*, he said, were both funny and sad, a "Johnny-come-lately nomenclature" that demonstrates little time for or interest in local knowledge. Instead, he proposed that if these species are to be named by science, it should be done in consultation with holders of traditional Hawaiian knowledge. Especially when it comes to endemic species, "Indigenous peoples should be given the priority and privilege to name things that come from their place." Naming ought to be an act of connection, or making relationship. In Larry's words: "When you name something, it makes it part of the family."

These days, snails and other species tend not to be named after the people who 'discovered' them. In fact, the *International Code of Zoological Nomenclature*—discussed further in the next chapter—includes an explicit rule prohibiting people from naming species after themselves. Nonetheless, it is still common for species to be named after other scientists—albeit with strangely Latinized versions of their names—often in an effort to acknowledge the labors of people who have dedicated their lives to the study or conservation of a particular corner of the biological world.[83]

While there are definitely connections being made through these contemporary taxonomic naming practices, there are also important possibilities for disconnection. As Larry noted, these naming practices can further entrench processes of alienation for Kānaka Maoli, both from the land and from dominant ways of knowing. Foreign names seem to describe an altogether foreign landscape. As Larry put it, "I don't know where they get these names, but you don't feel like you are part of that creation."

I'm genuinely unsure about what the right thing to do here is. Some scientists in Aotearoa/New Zealand and other places are calling for the revision of species names or at least new protocols for future names that acknowledge and work with Indigenous communities and naming practices.[84] Others have expressed concern over such suggestions, pointing both to the taxonomic confusion that would result from revising existing names and to the potential for a kind of appropriation in the shoehorning of Indigenous names and naming practices into the rigid systems of taxonomic nomenclature. One thing that is certain, though, is that there is a need for further dialogue on this topic. Larry's words open up the possibility for taxonomic research—related to snails and countless other species—to contribute to the broader work of (re)connection in a way that must not be overlooked.

Beyond naming, this work might also involve an opening-up of our understandings of the past. While figures like Gulick are frequently celebrated—and, as is the case around the world, Western men of this era are more likely than most to have their names attached to species—the reality is that Kānaka Maoli contributed significantly to the scientific endeavor in ways that are frequently ignored or reduced to "manual labor."[85] This kind of reevaluation of the historical record will be aided in Hawai'i by the incredible archive of Hawaiian-language newspapers, an archive that a growing group of Kanaka scholars is drawing on in just this way.[86]

Fascinatingly, in recent years we have seen the beginnings of important new dialogues between snail taxonomists and Kānaka Maoli, some of them grounded in the collected shells that have been the main focus of this chapter. Today, roughly 170 years after "land shell fever" began sweeping through the islands, staff at the Bishop Museum are working to give these shells a different afterlife. A key part of this effort lies in a large-scale digitization project, being led by Ken Hayes and Nori Yeung, that aims to make the shells, as well as collectors' field notes and journals, more readily usable and searchable by all.

The role of the Bishop Museum's malacology collection in current conservation efforts is discussed more fully in the next chapter. For now, however, it is important to note that this work also aims to open up the collection to conversations with cultural practitioners and the wider community. One of the key opportunities that the collection affords is to be a resource in efforts to interpret traditional Hawaiian moʻolelo and mele that refer to snails. Sometimes snails are discussed in particular places or in association with particular plants. In this way, various clues can be pieced together to make connections.

In one instance, Nori was contacted about a particular chant that referred to a snail with the name "hinihiniʻula" that was said to be found in Kohala, the northwestern part of the island of Hawaiʻi. This name—the term "ʻula" meaning scarlet or red—seemed to refer to the color of the snail. Drawing on the collection, Nori was able to scan through the shells of species found in that region and determine that there are no red-shelled snails to be found there. Digging a little deeper, though, she found an alternative, a species of the genus *Succinea*. The original description of the species notes its translucent shell, and beneath it a bright red body. This is one of the snail species that is still doing okay today. If you look hard enough in the hills of Kohala, its bright red form can still be found. But if this species were one of the many others that are no longer around, the collection and its records might very well now be the only way in which to make this connection.

Interestingly, however, this isn't the only possible interpretation for the name "hinihiniʻula." While the term "ʻula" does translate as red, it can also refer to something that is sacred or royal. For this reason, Pueo told me, the color red is sometimes referenced in referring to a rainbow. Indeed, the snail name "hinihiniʻula" is sometimes translated as "shell with beautiful rainbow colors."[87] However, when I asked Nori about the possibility that it might be another brightly colored snail species being sung about in this chant, she told me that no such snail is known from that region: all of the other species, both extinct and extant, are predominantly brown and white. For now, at least, it seems that the red-fleshed *Succinea* remains the best candidate.

This situation reminds us that making connections between the museum's shell collection and Hawaiian cultural practices and understandings is anything but straightforward. Building this kind of knowledge requires an intimate appreciation of the contours of snail lives and landscapes, as well as of language and cultural meanings. It will require collaboration between scientists and cultural practitioners, drawing these collected shells into dialogue with the extensive archive of 'ōlelo Hawai'i newspapers and other historical sources. All of this work is made that much more difficult by a long history of colonization and species loss that has impacted so significantly on the snails and the people of these islands.

These dialogues around snail shells are only just beginning at the Bishop Museum. As Nori explained: "It's really early for us to be able to do this. We're finally getting a sense of the knowledge that is locked in these collections. Once we have that opened, cultural practitioners can start accessing that information for themselves." In this way, this legacy of collecting might also begin to play a small role in strengthening connections between Kānaka Maoli, their stories, and the snails that they have shared these islands with for so long.

4 THE ANONYMOUS
Taxonomy and the Unknown Extinction Crisis

I spent hours poring over snail shells, opening one drawer after another, closely examining the diverse shapes, colors, and textures that each revealed. I have to admit that I didn't really know what I was looking at. I had then, and still have now, nothing like the trained eye of a taxonomist, honed to the craft of identifying and categorizing species. The longer I spent staring at the shells in the Bishop Museum's malacology collection, the more convinced I became that I lacked almost entirely the ability to discern the subtle differences of form that matter to the experts in telling apart two closely related species.

At this stage in my immersion into the worlds of snails I was even having trouble with one of the more straightforward taxonomic criteria: shell chirality. This term refers to the direction in which a snail's body, and so its shell, spirals. Holding a shell in front of me, and trying to determine if it was dextral or sinistral—that is, spiraling in a clockwise or a counterclockwise direction—my mind began to swim.

But at the same time there was something oddly meditative about my time among the shells: moving my eyes from one group to the next, drinking in the differences I could discern and not worrying too much about those that eluded me. If nothing else, I hoped that the carefully labeled specimens might allow me to mentally attach the image of a shell to some of the snail species that had come up in my research and conversations, to "put a shell to a name," so to speak.

My host was Nori Yeung, the curator of the collection. In addition to allowing me to spend some time quietly looking through the shells on my own, she had agreed to show me around and talk to me about her work. In particular, I wanted to speak to her about her taxonomic research, put simply, the effort to create a comprehensive inventory of snail life in the Hawaiian Islands. Drawing on this remarkable collection of shells and a broad array of other resources, Nori and her colleagues are working to further refine our understanding of just how many Hawaiian snail species there once were, as well as how many of them are still left.

In Hawai'i, 759 species of land snails have been recognized by taxonomists to date. But this number is subject to ongoing change, both as new species are discovered and as species formerly identified—some of them well over 100 years ago—are revised. In some cases, two species are found to actually be only one, or one species might need to be split into two or more.

This kind of taxonomic research is often taken for granted in conservation efforts, especially when we're dealing with vertebrate animals like mammals and birds where we now generally have a pretty good sense of who is who, and even of their life histories and distributions. When it comes to snails and many other invertebrates, however, there is so much that we don't know, including really basic things like whether these two snails are the same or different species. Without this kind of information, we can't even say whether conservation efforts are necessary, let alone how we ought to go about them.

These blank spots in our scientific knowledge are actually much larger than most of us assume. As soon as we venture out beyond the familiar terrain of the vertebrates, there are many more unknown species than there are known. The vast majority of these as-yet-unknown species are invertebrate animals, a catchall term for those without a backbone that includes everything from crabs, octopuses, and jellyfish to insects, spiders, worms, and, of course, snails.

In popular news media around the world, stories of recently discovered species are frequently reported as though new species are hard to come by. The reality is more complex. While, as we will see, identifying and formally describing a new species is an involved process, somewhere in the region of 10,000 such discoveries take place each year.[1] There are, in actuality, species unknown to science all around us, all the time.

The contemporary scientific effort to catalogue the diversity of life is generally traced back to the work of the eighteenth-century Swedish botanist Carl Linnaeus. In particular, it was he who refined and popularized the system of "binominal nomenclature"—that is, the system whereby each species is given a two-part Latin-esque name that includes its genus and species— in his book *Systema Naturae* and in so doing began a more unified effort to name and order the great "system of nature."

But despite more than two centuries of taxonomic work since Linnaeus, a great deal remains to be done. We don't have any hard numbers to rely on here: we don't know how many species there are on the planet, nor do we even know precisely how many have already been described, as there is no single master list that we can consult (although the Catalogue of Life, an international collaboration headquartered in the Netherlands, is working toward this goal). The best estimates of both figures, however, tell us that taxonomists have probably identified around 2 million of the roughly 10 million species of plants, animals, and fungi with which we share this planet.[2] This leaves roughly 8 million unknown species—the majority of which are thought to be invertebrate animals, especially insects.

These unknown species are not simply a puzzle for curious taxonomists. In our present time, many of these species—living kinds that have never quite managed to appear to us in the first place—are disappearing from the earth, forever. The fact that science has not yet named or described a species does not afford it any kind of special protection from extinction. Instead, as we will see, there is every reason to believe that these unknown species are being lost at least as quickly as those that we do know about—that is to say, at a staggering rate.

This complex situation burst onto the world stage in 2019 when one of the peak international bodies concerned with conservation, the Intergovernmental Science-Policy Platform on Biodiversity and Ecosystem Services (IPBES), announced that over a million species were at risk of extinction.[3] Skeptics immediately responded that the most comprehensive international list of threatened species, the IUCN Red List, at the time included "only" about 28,000 threatened species. The very significant gap between these two numbers is precisely what is at issue here. As previously discussed, the Red

List includes only those threatened species that not only have been described by science but have been the subject of detailed, ongoing studies. In contrast, the IPBES estimate of a million species was based on mathematical modelling in an effort to take account of the many species that we know little to nothing about, the vast majority of which we don't even have a name for.

There, among the shells at the museum, Nori introduced me to a tangible embodiment of this complex process of unknown loss. Alongside the rows and rows of cabinets of carefully catalogued shells, in dark corners and cupboards, other shells are waiting. Some of them, Nori told me, had been collected over 100 years ago; some were donated by private collectors, like those described in the previous chapter; others were gathered on museum expeditions. Either way, they arrived without the time or resources to adequately catalogue them. These shells now wait in cardboard boxes, old Mason jars, and an assortment of other containers. There are, without doubt, numerous "new" species—unknown to science—waiting to be described among these shells. We should expect, however, that when and if these species are one day described, most of them will already be extinct when this happens. Their shells will serve only as an announcement of a loss that was not known at the time of its occurrence.

The disappearance of unknown species is a pervasive and largely undiscussed feature of our current period of massive global biodiversity loss, one that, as we will see, might be understood to constitute its own unknown extinction crisis. In this context, this chapter is an effort to grapple with the unknown. It explores how the islands' snails come to be known to science and to be conserved, and the challenges and limitations of these processes. In this way, this chapter aims to cast a light on the threat of ignorance and indifference that surrounds the loss of snails and a host of other invertebrate species today.

There are numerous methods of classifying the diversity of life on this planet. All communities use their own particular taxonomic systems and "common names" that align with and diverge from those of scientific taxonomy in different ways. In the previous chapter we encountered some Kanaka Maoli names for snails and ways of understanding and storying their place in the wider world. Alongside these differing human modes of classification, it is also important to acknowledge that other living beings make their own "taxonomic" distinctions among themselves, deciding in a host of ways who is and isn't one of their kind, or one of a variety of other relevant kinds who is a meaningful part of their world. As we saw in chapter 1, among their various modes of making such distinctions, snails use slime trails to determine who might be a potential mate, companion, or meal.

As such, my discussion in this chapter of taxonomic science and its domains of classification, of the named and the unnamed, should not be taken to represent an account of anything like the absolute or universal limits of such possibilities. Rather, it is an account of one particular set of practices, at one particular point in time, albeit, as we will see, a set of practices that represents a highly consequential mode of understanding and ordering life. As I've gotten to know this taxonomic work a little more, I have discovered that it provides a fascinating window into our biological world, one that helps us to appreciate in new ways the intricate patterns and processes of coming into being, solidification, and passing away of living kinds.

A TREASURE MAP

Housed on the top floor of Konia Hall, the Bishop Museum's Division of Malacology has a strangely bunker-like feel to it. The dim lighting, the solid concrete walls, and the loud, steady hum of the air conditioning create a feeling of secure enclosure. It is a facility designed to store precious things, insulated from the light and temperature fluctuations of the outside world. Dominating the entire space, from the moment you open the door, are the white and gray metal cabinets that contain most of those precious things: the shells. Each row comprises about 40 cabinets, stacked two high, each of which contains 22 shallow drawers, which are in turn divided up by small cardboard trays of various sizes. Each of these trays is home to a single lot of shells of a particular snail species gathered at a particular time and place.

This incredible collection is tied intimately to the lives of two men, Charles Montague Cooke Jr. (1874–1948) and Yoshio Kondo (1910–1990), who together oversaw its creation and care for roughly the first 80 years of its life. Cooke was born in Honolulu and after being educated at Yale

University returned to the islands where, in 1907, he took up the role as the Bishop Museum's first Curator of Pulmonata (air-breathing snails and slugs).[4] Over the next 4 decades, Cooke became a highly respected malacologist and transformed the museum into a leading site for research on Pacific mollusks. For the last decade and a half of his career, Cooke worked closely with Yoshio Kondo. Kondo initially served as an assistant in the collection, and studied biology in Hawai'i before earning his PhD at Harvard University. After Cooke's death, Kondo took over as head of the division, and held this position for about 30 years, until his retirement in 1980.[5]

Standing in the malacology collection today, the legacy of these two men is readily apparent. It is there in the photographs of them on the walls and the lingering traces of Cooke's pipe tobacco in the shell cabinets. It is there in the simple sign that hangs above the entry to the small Kondo Library, as well as the filing cabinets filled with the manuscripts, sketches, and hand-drawn maps that they produced through decades of fieldwork and research. It is there also in the hundreds of glass jars of preserved snail bodies that line one large wall, filling a series of timber cabinets that were custom made for them at some point. Nori explained to me that while most malacologists of the period simply discarded them, "Cooke had the foresight to keep all of the bodies that he collected. . . . Most of them are preserved in pineapple alcohol that he distilled, personally."[6]

But, of course, the most significant legacy—in terms of both its scientific value and its sheer size—is the shell collection itself. As I explored the cabinets, the beauty of these shells drew me in, but so too did the tiny, immaculate, handwritten numbers on many of the individual specimens. When I asked Nori about the careful handwriting she replied, "Oh, that's Cooke." The most significant period of acquisition occurred under his leadership, with some substantial new additions in Kondo's time.

Many of these shells were personally collected by these two men and their colleagues. In the first half of the twentieth century, in particular, the museum oversaw a significant program of expeditions and smaller collecting trips in Hawai'i and the wider Pacific. Cooke led one particularly important trip, the Mangarevan Expedition, which in 1934 took a small team of scientists on a 6-month, 14,000-kilometer trip to islands in the

Achatinella shells in the Bishop Museum collection, labeled by Cooke.

Society, Gambier, and Austral groups.[7] It was on this trip that the young Kondo, who was serving as the ship's chief engineer, had his first encounter with malacology.[8] As we have seen, it stuck, and Kondo went on to a life of snail-focused research, including launching many of his own collecting trips around the Pacific.

But despite these extensive collecting efforts over decades, a significant number of the shells now held at the museum were not purposefully gathered by its scientists. Instead, they were acquired by the museum from the private collections produced by the army of naturalists and shell enthusiasts whose activities were described in the previous chapter. These collections—sold, donated, or bequeathed—frequently made their way to museums. Later in his life, having finished his research, even the enthusiastic John Thomas Gulick parted with his shells, dividing his collection of well over 40,000 into 20 sets to be sold or donated to museums. Cooke, during his tenure, secured 11 of these sets for the Bishop Museum.[9]

By the time of Cooke's death in 1948, roughly 5 million specimens had been added to the collection. Today, it is not a lot bigger than it was then. With a total of roughly 6 million terrestrial, marine, and freshwater Pacific island snail shells, it is the most comprehensive collection of its kind in the world. Although shells come from 28 island groups across the Pacific, there is a distinct focus: roughly 40% of the entire collection is from the endemic land snails of the Hawaiian Islands.[10]

In the decades that followed Kondo's retirement, both the collection and the study of Hawaiian land snail taxonomy experienced some significant ups and downs. Carl Christensen took over as curator of the collection in 1980. Having recently completed his PhD, he had arrived at the museum 2 years earlier as an unfunded researcher, taking on work with archaeologists identifying the remains of long-dead marine mollusks. When Kondo retired, Carl jumped at the opportunity to take up one of the very few funded positions for a malacologist in the islands, in charge of a collection that he had spent his summers working in as an intern during high school.

But this was a less than ideal period in which to be advocating for snail science in the islands. During the 1970s and 1980s, there was very little interest in or resources for this work. In fact, Carl told me that during this period "Mike Hadfield and I were the only people [in the islands] who had any professional interest in Hawaiian land snails." At the same time, the shell collection itself was beginning to suffer from a lack of investment in its maintenance. Working closely with the shells, Carl became increasingly worried by the powdery white substance that he found forming on a growing number of them. While some shells were just covered in a light dusting, others had become badly scarred and even begun to disintegrate.

Carl determined that the situation was caused by Byne's disease, a condition named for the British naturalist Loftus St. George Byne, who published a description of it in 1899. Byne's analysis of his eponymous disease, however, was not a particularly helpful one. As Paul Callomon, the manager of the malacology collections of the Academy of Natural Sciences in Philadelphia, has put it: "Most of his conclusions were wrong—he thought the problem was caused by bacteria—and the treatments he prescribed were either useless or actually dangerous, but the name stuck."[11] Subsequent research has shown that there is no "disease" here. Instead, this damage is the result of acetic and formic acids that are naturally produced by the breakdown of wood, cardboard, paper, and other cellulose-based materials. These acids react with calcium carbonate shells, creating compound hydrated calcium acetate and calcium formate salts. It is these salts that do all the damage and that are visible as a white powder or efflorescence on affected shells.

Before Carl had a chance to really address this situation, however, his position was terminated. In June 1985, as part of the turnover process, he wrote a memo to his direct supervisor, Allen Allison, the head of Zoology, in which he described the problem and warned of its potential impacts on the collection. Allison hired a consultant to analyze the situation and began the process of improving the environment in which the shells were stored, installing air conditioning, and sealing up the space. Lowering the temperature and humidity and reducing their fluctuations in this way slows the chemical processes that drive Byne's disease, a particularly important consideration for a large shell collection kept in Hawai'i's tropical climate. But the majority of the shells were at this point still housed in timber cabinets and cardboard boxes, and with no funding available to replace them, the problem was going to persist into the future.

The termination of Carl's position as curator of the collection came only 5 years after he had taken it up. Almost a quarter of the institution's scientific staff were fired at this time as part of a larger restructuring that took place with a change of senior leadership at the museum that is widely seen as having deemphasized research and collections in favor of public engagement.[12] For a while Carl looked for other malacological work in the islands and elsewhere, but eventually had to accept that there was none to be found. At the age of 39, he made an about-face in his career and enrolled in law school. For the next 5 years the malacology collection was "orphaned," without a curatorial head. During this period, Regie Kawamoto, a marine collection technician with a half-time position at the museum funded privately by the malacologist E. Alison Kay, continued to look after its day-to-day care.

In 1990, Robert Cowie was hired as the collection's new curator, with a little help from its first curator, Charles Montague Cooke Jr., or, rather, from his family. The initial 2 years of funding for Rob's position were provided by the Cooke Foundation, established in 1920 by Monte's mother, Anna. On taking up the position, Rob immediately set about applying for grants to make the storage upgrades needed to ensure the long-term health of the shells. A couple of years later, with substantial funding from the National Science Foundation (NSF), he was able to purchase the metal cabinets that now fill this space. With the help of grant-funded staff and some volunteers, he took up the mammoth task of transferring the land shell collection from its existing storage to its new home, including archiving all of the old labels and notes. As Rob succinctly put it to me: "It took forever. That was the first four years of my job at the museum."

In his down time from this project, however, Rob managed to achieve another important milestone. Trawling through hundreds of published articles written by biologists and naturalists over the previous 200 years, he compiled the first comprehensive and rigorous catalogue of the native land and fresh-water mollusks of Hawai'i. Drawing on significant input from Carl and Neal Evenhuis, a colleague at the museum, this catalogue provided a baseline for all subsequent taxonomic work in this area.[13] As we will see, taxonomic discovery and revision continue to this day, but it is on the basis of this catalogue, with a handful of species described since its publication, that we can today say that there are 759 currently recognized species of native Hawaiian land snails.[14]

Rob explained to me that his effort to rehouse the collection, combined with the research to develop this catalogue, allowed him to develop an in-depth sense of the scope and value of the collection and drove his effort to "put it back on the map" for malacologists all over the world: to expand the loan of materials, to get more visiting researchers involved in the collection, and ultimately to ensure that both the collection and Hawai'i's snails were a visible and active site of research interest.

In 2001, despite his commitment to the collection, the prospects of a more stable and better-funded position led Rob to move to the University of Hawai'i to take up his current position. This incredible archive of snail life was again without a curator. Regie continued in her half-time role, taking care of the collection, cataloguing new arrivals, and filling loan requests. Things remained like that for over a decade, until 2015, when Nori stepped into the role, becoming the collection's fifth curator in its more than 100-year history. As with the struggle to conserve Hawai'i's living snails, the work of maintaining the health of this archive of their past is ongoing. Among her many other tasks, Nori is currently overseeing a program in which almost 117,000 glass vials of minute shells are having the cotton wool that was used as stoppers replaced with a synthetic substitute.

This latest part of the ongoing effort to protect the collection from Byne's disease has also been funded by the NSF. It is one element of a larger project led by Nori and her partner and colleague Ken Hayes, who is the Director of the Bishop Museum's Pacific Center for Molecular Biodiversity. Together, they are working to both conserve the collection and enhance its accessibility, including through the digitization efforts mentioned in the previous chapter. They have utilized the opportunity afforded by this funding, and the extensive work required by these tasks, to bring student and volunteer interns back into the collection, training them in taxonomy, museum curation, conservation biology, and more. In this effort, Nori told me, they see themselves to be "following the training footsteps of Kondo," cultivating the next generation of scientists who will help to take care of this collection and the snails whose lives it documents.

On first encountering this incredible collection of shells, I assumed—like most people, I think—that it was something of a dusty relic from bygone days. The reality, however, is much more interesting. Although the heyday of shell collecting in Hawai'i has long passed, this collection at the Bishop Museum continues to grow, albeit much more slowly. This growth is an integral part of the active life of this collection and the important work that it does in helping us both to better understand and to conserve the islands' remaining species. In fact, a key part of what is revealed by getting to know this collection is that understanding and conserving snails are not at all separate pursuits—each of them requires the other in profound, but not always immediately obvious, ways.

Over roughly the past decade, Nori and Ken's work has focused on tapping into the collection to guide a program of extensive surveys across the Hawaiian Islands in search of lost snails. Ken explained to me that the seed for the project lay in his first years in Hawai'i, when he and subsequently Nori were graduate student assistants working with Rob at the University of Hawai'i. During this time Rob and Ken were conducting a range of surveys to develop a better understanding of the invasive freshwater apple snails and the many other introduced gastropods that had made their way into the Hawaiian Islands, some of which have significant impacts for agriculture and human health.

But as they spent time in the field, they were coming across more and more snails that hardly anyone could identify. Ken started to realize that many of these snails were in fact native species, just rarely seen ones. Some of them, he told me, were species that had been spotted for the first and last time "at their original description, maybe in the 1910s or 1920s." Many of these species had been presumed extinct, or at the very least had largely dropped off the malacological radar. In an effort to identify these snails, Ken turned to the Bishop Museum collection and began to wonder about how many other species were still out there, perhaps just hanging on.

Since 2010, Nori, Ken, and their collaborators have surveyed roughly 1,000 sites across all of the main Hawaiian Islands, the most extensive survey of native land snails ever conducted here. This survey work was initially conducted from Rob's lab with NSF funding, but later moved over to the Bishop Museum—along with the growing collection of shells it was producing—when Nori took up her current position there. Along the way, many other people have been enrolled into this work, including Dave Sischo and the Snail Extinction Prevention Program, as well as other agencies and organizations that spend time in likely snail habitat, with key personnel being provided with taxonomic training where needed to aid in identification.

The collection has guided all of this work. The shells themselves are vital, but so too are the journals, maps, and records kept by the more scientifically inclined collectors of the past, from the mid-nineteenth century to today. Some of these people provided detailed sketches of hills and valleys, labeling exactly what was found where. These resources now guide the way to possible remaining populations. As Nori put it: "This collection has been like my treasure map: without it we wouldn't know how much has been lost, but also how much can still be saved."

But the collection offers other insights, too, especially for those who know how to read its materials in combination. In some cases, they have been able to make comparisons between species to hypothesize about their past relative abundance based on the number of shells collected for each (controlling for important variables). In other cases, they have compared historical observations about abundance in field journals with contemporary survey data. Based on the various sites in which a species was collected, it is sometimes also possible to infer past distribution: it was Cooke's records on the distributions of *Auriculella diaphana*, published in 1912–1914, that Brenden made use of to determine the roughly 99.9% range contraction of the species that I discussed in chapter 2.

This current work all takes place in the context of a long period of relative inactivity in land snail research in Hawai'i. While individuals within the malacological and conservation communities were aware that species were disappearing, and some published articles sought to draw attention to the situation, for a couple of decades from the 1960s very little new research was conducted to study or record that decline.[15] Citing the account of Alan Solem from the Field Museum in Chicago, Nori and Ken have referred to this as a period of "malacological silence" in Hawai'i.[16] As we have seen, this situation was compounded by a lack of support and resources, including at the Bishop Museum.

While things began to pick up again in the 1980s, with work by Mike Hadfield, Alan Hart, and others to both publicize and conserve the genus *Achatinella*, the vast majority of Hawai'i's land snails continued to be largely ignored as many of them quietly disappeared. Over the past decade, Nori and Ken have worked with the SEPP team and others to shift this *Achatinella*-centrism. While a focus on the larger and more charismatic snails—relatively speaking, of course—did important work in earlier decades, today they believe it has become a liability. As such, they insist on the importance of studying and conserving *all* of Hawai'i's snails—no matter how drab and unremarkable some of them might be in appearance. Through their survey work, Ken explained, "we've rediscovered dozens and dozens of species that were thought extinct at some time."

As important as these rediscoveries are, there is no denying that they take place within an overriding and indeed overwhelming story of loss. Take the family Amastridae, for example. Through Nori and Ken's work, we have gone from knowing about 15 extant species in 2015 to 23 species today. Eight species thought lost forever are still with us. This is undoubtedly good news, but the fact remains that this family was once by far the largest and most diverse in the islands, including 325 recognized species—in other words, we have still lost roughly 300 other species of amastrids to date.

A similar pattern has been repeated, again and again, throughout the Hawaiian Islands to produce our current predicament. Through their comprehensive field research, Nori and Ken have documented the loss of roughly 450 endemic land snail species, and the precarious state of most of those that remain. As noted in the introduction to this book, their latest figures indicate that of the roughly 300 species that are left across all of the Hawaiian snail families, a total of 11 can be categorized as "stable."[17] In this context, while I want to celebrate the eight additional species of amastrids, and the handful of others thought lost that are not yet, spending too long on this good news feels a bit like scratching around in the rubble for a glimmer of hope.

TRIAGE TAXONOMY FOR AN INVERTEBRATE CRISIS

There are several ways to extract DNA from a snail. In the past, these processes have been a lot more invasive and frequently lethal, involving snipping tissue samples from a snail's foot or the use of their whole bodies. This is a less than ideal situation for any snail, but it raises particular problems when working with a threatened species. These days, as the genetic techniques have become more sensitive, other options have opened up. For the larger snails like the *Achatinella* species, the procedure is now often as simple as using a Q-tip to swab mucus from their foot. Smaller species, which don't produce enough mucus for this to be effective, can now be placed so they can crawl over an FTA card, a piece of chemically treated paper that retains and preserves the DNA in their mucus.[18] For the truly tiny snails, though, those under about 5 millimeters in length, even this noninvasive approach isn't an option: the FTA card contains salts that can be harmful or even lethal to a snail of this size.

I learned about these techniques chatting with Ken about the work that he does in the museum's Pacific Center for Molecular Biodiversity. Located in the building next door to the malacology collection, in many ways this center feels like it is a world away. In place of the rows of cabinets containing shells and preserved snail bodies that fill the Division of Malacology, the center has a distinctly high-tech feel, benches lined with an assortment of laboratory equipment and computers for molecular analysis and tissue cryopreservation.

But despite my initial impressions, I quickly learned that the work carried out in these two parts of the museum is thoroughly integrated. Each needs the other; both play vital roles in the work of snail taxonomy, in efforts to identify new species and to refine our understanding of those that have been previously described. In the past, many new species were declared solely on the basis of distribution and shell morphology (physical characteristics). Indeed, for a long time the study of snail taxonomy was so shell-focused that it was largely understood to be the province of conchologists, those whose sole focus is these calciferous components. On this basis, species went in and out of (taxonomic) existence as their validity was revised by subsequent analysis. In the nineteenth century, the *Achatinella* tree snails were a subject of particular interest and thus contention in this regard—with some species being proposed, disposed of, and then restored, all in the space of a few decades.[19]

This focus on shells began to shift internationally in the late nineteenth century, as it was increasingly realized that these hard components did not necessarily provide an accurate picture of how snails should be organized into species and larger groups. As the malacologist Robert Cameron has put it, "internal anatomy told different stories."[20] Cooke was perhaps the first to pay real attention to the fleshy parts of Hawai'i's snails in his taxonomic work, and certainly to preserve them, using these bodies to compare reproductive systems and other anatomical features for important taxonomic clues.[21]

Today, the tools of molecular biology have joined this lineup. DNA samples are taken from snails, and key gene sequences are analyzed and compared.

The degree of genetic similarity between organisms helps to determine whether or not they are members of the same species, while the genetic similarity between species in turn provides information about how closely related they are to one another in evolutionary terms (phylogenetically). In addition to sampling freshly collected snails in the ways I've just described, Cooke's pineapple alcohol–preserved bodies are proving to be an invaluable resource yet again, yielding usable DNA from many species that are now extinct.

This taxonomy is a work of care, requiring meticulous attention to detail. Importantly, new approaches to taxonomy have not simply superseded the old. Instead, they have been layered over one another to produce an integrative approach that is now considered by many to be a requirement for good taxonomic work.[22] When it comes to snails, it is an approach grounded in the understanding that no single part of the organism—not its shell, its fleshy anatomy, or its DNA—can provide simple and definitive answers. In each case, similarities and differences might reflect phylogenetic relationships, or they may be the result of convergent evolution. For example, similar shell structures may be the result of adaptation to a common lifestyle or predator. Consider, for example, the very similar, yet unrelated, shells of Hawai'i's *Achatinella* snails and those of the *Liguus* snails of Florida, two groups of tree snails whose shells seem to have converged on roughly the same form as a result of their similar lifestyles.[23] Genetic similarities, Ken and Nori explained, might be produced in the same way. If, for example, you're looking at the genes associated with temperature tolerance, they might accumulate mutations in a way that makes two species that have undergone similar selection pressures look more related than they are. Examining multiple gene sequences can help to address this issue, but it cannot overcome it altogether.

When, through a combination of these various strands of information, a new species is discovered, it formally enters the space of scientific knowledge through the act of being described in a published article. This practice dates back to Linnaeus and the birth of modern taxonomy, but over the centuries it has been increasingly formalized. In the case of animals, much of this process is now governed by the *International Code of Zoological Nomenclature*. There is, however, still a fair bit of freedom about what information a description includes and where it is published. These days, a good description provides information on anatomy and molecular biology, the variation found among individuals, as well as their lifecycle, habitat, and distribution, often alongside photographs and other relevant illustrations. In short, as Ken summed it up, a good description includes "everything that someone else would need to discern whether or not what they find when they're out in the field is the same thing you're describing." Of course, the description also includes a name for the species.

Alongside this publication, describing a new species also involves the act of depositing a "type specimen" into a museum or another secure archive. This type specimen, or holotype, plays an essential role in the modern taxonomic project. It is the individual organism—in the case of snails, generally a shell and perhaps also a preserved body—to which, as the historian of science Lorraine Daston has put it, "the original description and name is anchored," to be consulted by subsequent generations of taxonomists in the future.[24]

Talking to Nori, Ken, and a range of other taxonomists, I slowly came to appreciate just how complex this work is. Good taxonomy cannot be conducted from an armchair; it requires the drawing together of multiple strands of information, techniques, methods, and expertise. Nori and Ken's work takes them into the field, the laboratory, and not only their own but multiple other museum collections all over the world. I stumbled upon firsthand evidence of this painstaking research at the Natural History Museum in London when I was lucky enough to be able to go behind the scenes to see their collection of snail type specimens. Each time I came across a Hawaiian species, there was a little note tucked in with the shells, like a strange, very brief, postcard from Hawai'i: "Examined/photographed by Norine Yeung during visit 10–14 Mar 2014."

Alongside this close attention to detail, taxonomy is also a work of care in the sense that it is animated and guided by an understanding that accurately cataloguing this diversity is now essential to the ongoing survival of those species that remain. As Ken explained to me, snails and other invertebrates present conservationists with a really difficult situation. While we know a great deal about many of the birds and mammals now facing extinction, this simply is not the case for these others.

> We don't know how to save them because we can't even name most of them. We don't even *have* a name for them, we don't know if that's the same species as this one here. And if we can't name it, we can't tell you anything about its biology. We can't tell you how many offspring it has per year, we don't know how it mates, we don't know what it eats, we know almost nothing about them.

The designation and naming of a species is an essential part of the work of care as it is practiced in contemporary biodiversity conservation. As the philosopher Joshua Trey Barnett has noted, taxonomic work brings species into the world in some sense, as distinct, concrete entities: "The act of naming delivers 'species,' which strictly speaking cannot be observed, over to us as something we can consciously consider, think about, write on, and care for."[25] While this is true in some sense for all species, it is particularly the case for many invertebrates among whom distinctions between species are very far from obvious, and certainly not readily visible. Instead, as we have seen, these species must be carefully crafted at the intersection of diverse practices, from conchology to genetics.

Of course, the relationship between description and conservation is not straightforward. Description is neither a sufficient nor a necessary condition for conservation: many described species are not conserved, and some species are conserved without being described (e.g., if they happen to reside within a protected area). In fact, in some cases being described can increase the threats to a species by making it more noticeable to collectors and others. And yet, in a variety of ways, taxonomic names have today become a precondition for vital forms of visibility and care, including for the kind of basic research that Ken is describing, as well as for the allocation of conservation funding under the species-centric regimes like the US Endangered Species Act that now dominate conservation efforts in many parts of the world.

Taxonomy takes on a very particular form when it comes to organisms like Hawai'i's snails that are highly diverse and relatively understudied, but also rapidly disappearing. In these contexts, the taxonomy can't simply be assumed—it doesn't sit quietly in the background, largely settled, rearing its head every now and again when, for example, a species is reclassified as a subspecies. Instead, for snails and many other invertebrates, taxonomic work takes place in ongoing dialogue with the efforts of conservationists. This situation was driven home for me when Ken referred to their work as "triage taxonomy": they are identifying species in an effort to save them, focusing on species that there is still time to help, describing them so that they might be added to the SEPP list of those in need, and perhaps one day even officially listed as threatened.

Malacological collections like the one at the Bishop Museum— collections of shells, bodies, DNA samples, documents, and more—are essential to this work. But not just any collections: they need to be living, well-managed, and well-resourced collections, not orphaned collections left to deteriorate and accumulate dust. They also need to be located in place, connected to the communities that use them, in this case including the Kanaka Maoli cultural practitioners discussed in the previous chapter, and the conservationists discussed in this one.

Under Nori's leadership, the museum's collection continues to grow. Indeed, Nori and Ken insist that it must do so, as a record for the future of both our shifting understandings and the changing distributions and existence of species. Of course, collection practices have changed with new understandings and technologies, and nothing like the scale of past collecting—either of Cooke's day or of the nineteenth century—is required today, even if it were still possible. But some collecting remains essential, nonetheless. Some snails are still brought back to the museum to be compared closely with other shells in this or another collection, or to be subjected to DNA analysis in the lab. And then they need to be carefully stored— which unavoidably involves their deaths—so that they can become a part of this collection and its rich, continually unfolding, story of snail life in the islands.

Talking to the staff at the Bishop Museum I got my first real sense of the "invertebrate bias" at the heart of our current biodiversity crisis. It is difficult to adequately express just how little we know about the crawling, buzzing, fluttering, creeping, and, of course, sliming world of invertebrate creatures around us, as well as the sheer voluminous diversity of their species. While there remains a great deal to be learned about all sorts of facets of our living world, from the deep oceans to the intricacies of cells and genes, many of the tiny critters in our backyards—and certainly the ones that reside in less convenient locations—also represent huge gaps in the collective knowledge project of modern science.

Invertebrates make up the vast majority of the animal kingdom, probably somewhere in the order of 99%. For every species of tiger, whale, owl, or frog, there are roughly 100 different kinds of crabs, ants, worms, jellyfish, and snails. These species are also vital contributors to the health of our planet in innumerable ways. While it is still a relatively understudied area, more and more research is highlighting the significant roles of these species as decomposers, pollinators, seed dispersers, nutrient cyclers, and more, alongside the breakdown of these roles as a result of invertebrate declines.[26] (Importantly, however, we need to find ways to care for these species even if they don't, or no longer, play these kinds of clear ecological roles, as discussed in chapter 2.)

The lack of scientific knowledge about these species is both a result of and reinforced by a general bias toward research on vertebrates. If we divided up the world's zoologists along the lines of their specializations, there would be roughly 100 times as many working on each of the vertebrate species as there was for each invertebrate.[27] What's more, the nature of their research is often different. While taxonomy and basic biology still occupy much of the time of invertebrate specialists, those focused on mammals and birds are far more likely to be building fuller pictures of their ecology, behavior, and conservation status.

This situation is about much more than the depth with which we understand the world around us. It is also, fundamentally, about our capacity to protect and hold onto that world. Having not even been described, the majority of the planet's invertebrates are largely invisible within the modern conservation picture. But even those that have been described tend not to be particularly visible. Most of them lack the data to allow their conservation status to be assessed. One study found that while 90% of the mammals, birds, and amphibians had been evaluated in this way, among the *described* mollusks the figure was 3%. And mollusks are one of the best-studied invertebrate groups; the figure for insects is closer to 0.08%.[28] This means that, as Nori summed it up, the conservation status of fewer than 1% of the world's invertebrate species has been assessed. For all of the others, we just don't know enough to say how they're doing.

The simple reality is that systems like the IUCN Red List were not designed for invertebrates. Mollusks had to wait until 1983, about 20 years after its establishment, to be included at all. They, like the other invertebrates, plants, and fungi—to say nothing of the bacteria and archaea that are far more difficult to categorize into species—are, as one eminent group of scientists succinctly put it, "grossly underrepresented" in its listing.[29] A large part of the issue is that Red List assessment requires detailed information on species range and demography, with surveys repeated over time to show trends. In the vast majority of cases, this is simply information that we do not have for invertebrates. Even among listed species, our knowledge of the invertebrates is generally much thinner: as compared with the vertebrates, roughly 12 times fewer conservation papers are published on the average listed invertebrate species.[30]

This bias remains apparent in numerous other parts of the IUCN's operations, for example, in the formation of Species Survival Commission Specialist Groups. These are panels of experts who focus their attention on a given taxon with the aim of aiding its conservation. While there are 73 such groups for the vertebrates, there are a mere 12 groups to look after the other 99% of the animal kingdom.[31] To put this in some context, albeit with some rather extreme examples, a single specialist group handles the entire phylum Mollusca, which comprises over 100,000 species of snails, slugs, octopuses, squids, and more, while a similarly sized group works on the African elephant, with a separate group dedicated to the Asian elephant.[32] While I certainly would not begrudge the elephants their specialists, there is clearly an imbalance at work here.

Of course, this kind of bias is not an IUCN-specific problem. The simple fact is that invertebrates very rarely receive anything like parity in research attention, funding, or public interest with their vertebrate counterparts. We see this at every level of governments, in the priorities of NGOs, at the zoo, in children's books, and in every other level of our biodiversity education. Ultimately, this situation also means that even if an invertebrate species is able to jump all of the hurdles to be formally listed as endangered, it is that much less likely to receive the kind of public support and interest that leads to funding and conservation success.

Invertebrates suffer from a kind of threefold ignorance here: we don't know most species; those we do know we often don't have the data to list as threatened; and even if they are listed, we often don't know enough about them to really conserve them. Underlying all of this is the fact that the public, in general terms, doesn't seem to really care about most invertebrates, whether they're named, described, listed, or whatever else. Indifference and ignorance feed into one another, reinforcing each other, and driving the relentless loss of invertebrate diversity around the world. And yet, as best we can tell, invertebrate species are disappearing at a staggering rate, albeit in a way that is, as biologist Nico Eisenhauer and colleagues have put it, "quiet and underappreciated."[33]

The state of our current knowledge with regard to invertebrates reminds us that for a species to be "known" in a meaningful sense requires much more than it simply having been described and recognized as a species. The division between the known and the unknown is not a black and white one, but a space composed of many gradations of gray. As the biologist Alain Dubois has noted: "it would be misleading to consider that these 1.75 million 'named' species are 'known to science' [now closer to two million]. Actually, many of them (in an unknown proportion) have only been the subject of a single scientific publication, providing the original description of the type specimen(s), and are scarcely more than *mere nomina on lists*."[34]

Clearly, our current time calls for a period of intense and passionate immersion in the lives of the many little creatures we share this planet with. Despite the efforts of many taxonomists, however, there is little evidence that we're getting the job done in time. Amid a long list of other difficult insights that Ken shared with me, one of the most sobering was that at the current rate of taxonomic progress it will take roughly another 500 years to describe all of the world's invertebrate species. At current rates of extinction, hundreds of thousands, perhaps millions, more of them will be gone by then.[35]

UNKNOWN EXTINCTIONS

Not all of the snail shells that have found a home at the Bishop Museum have made their way into its carefully catalogued drawers. As I noted earlier, over the more than a century that this collection has been growing, a variety of shells have arrived without there being the time or resources to process them. And so now they sit, in an assortment of containers, waiting to be examined, identified, and catalogued. Alongside these uncatalogued shells, the malacology collection also includes a huge number of specimens that, while they have technically been catalogued, are in need of closer attention. Some of them were labeled with a tentative species name at the time of arrival; others might have been identified only down to the family or genus level. Either way, no one is quite sure exactly what, or who, they are.

In all, Nori estimates that there are around 3 million shells in the collection that either are uncatalogued or are missing important information. As a result, there are undoubtedly many species that are unknown to science waiting patiently to be described. This means that in addition to being an invaluable guide to locating rare or lost snail species in the field, the collection is itself also an active site for exploration and discovery.

As we have seen, the process of describing a new species requires considerable time and research. It is perhaps unsurprising, therefore, that there tends to be a delay between initial collection and formal description. Very often, this is a significant delay. Hawai'i's snails are far from unique in this regard. Around the world, the average period between the collection of the first specimen of a new species and its official description is 21 years.[36] As a result, one recent study has estimated that, internationally, there may be as many as half a million undescribed species already in museum collections.[37]

This particular collection at the Bishop Museum has played a significant role in the discovery of many new snail species. At the time of my last visit, Nori, Ken, and colleagues were working on the description of a species from shells collected by Cooke in 1924. Cooke suspected at the time that this was a new species, but he simply ran out of time to describe it. As Ken explained to me, about 50% of the specimens collected in Cooke's day were new to science. Today, with many more species described, he estimates the figure is more like 10%, but this still creates an incredible demand on time and resources.

One particularly rich group of snails thought to be awaiting description in the Bishop Museum is the Endodontidae. This family of tiny snails is spread across the islands of the Pacific Ocean; in fact, it is thought by some to have once been the most diverse of the larger families in the region.[38] In each locale where these snails are found, new species have evolved—sometimes a great many of them. When Alan Solem found his way into the aisles of the Bishop Museum's collection in the 1970s, he is said to have been astounded. Solem was, during his lifetime, the foremost authority on the endodontids, having authored two major volumes on them across the Pacific, describing numerous new species. At the Bishop Museum, he reported that there was another lifetime of work to be done, with shells from perhaps as many as 300 more undescribed Hawaiian species in the family awaiting the attention of a careful taxonomist.[39] Today, half a century later, these shells are still waiting.

When I asked Nori and Ken why no one had taken up this descriptive work, their response illuminated many of the complex issues of invertebrate bias, gastropod taxonomy, conservation, and extinction that are at the heart of this chapter. Part of the problem, they explained, is that all that we have of these "would be species" is their shells. As such, we're missing helpful morphological data, but also the molecular genetics that might have enabled some of the easier, quicker distinctions to be made. While Solem, drawing on over 30 years of experience working on the taxonomy of this family, might have been able to distinguish between species on the basis of their shells, as Ken summed it up: "we don't have anyone like him anymore." In order to really get to the bottom of this family in Hawai'i, Ken speculated that someone might need to devote 10 years to the puzzle, learning to really see and understand these snails in their regional context, so as to make the kinds of distinctions needed to identify species. But who will take this on? Who has the time and resources? What's more, Ken and Nori explained, for an enthusiastic junior researcher, this kind of work would likely be a career dead-end as there are so few positions today for this old-school approach to taxonomy.

Plate 1
Achatinella lila, Oʻahu. The nine remaining species of this genus are all listed under the Endangered Species Act. Photo by David R. Sischo.

Plate 2

Achatinella lila, Oʻahu. Once found in the Koʻolau Range, the last free-living population disappeared several years ago. Photo by David R. Sischo.

Plate 3
Partulina proxima, Molokai. Photo by David R. Sischo.

Plate 4

Achatinella mustelina, Oʻahu. This species is the central focus of the Army's conservation efforts that began in Mākua Valley. Photo by David R. Sischo.

Plate 5
Achatinella sowerbyana, Oʻahu. Photo by David R. Sischo.

Plate 6
Amastra intermedia, Oʻahu. This species was reduced to a mere handful of individuals but survived in captivity. Photo by David R. Sischo.

Plate 7

Auriculella turritella, Oʻahu. A species with a beautiful, conical shell. Photo by Kenneth Hayes and Norine Yeung.

Plate 8

A tiny snail of the genus *Punctum*, O'ahu. Photo by Kenneth Hayes and Norine Yeung.

Plate 9

Endodonta christenseni, Nihoa. Described in 2020 by Nori and Ken, and named in honor of Carl Christensen for his considerable contributions to snail conservation and taxonomy. Photo by David R. Sischo.

Plate 10

Achatinella decipiens, Oʻahu. A group of snails estivating together. Photo by David R. Sischo.

Plate 11

Tiny snails of the genus *Tornatellides*, Oʻahu (subfamily Tornatellidinae). Perhaps similar in size to those first individuals to arrive in these islands by bird. Photo by Kenneth Hayes and Norine Yeung.

Plate 12

Newcombia cumingi, Maui. While this island has many rare snail species, this is the only one that has formal listing under the Endangered Species Act. Photo by David R. Sischo.

Plate 13

Laminella aspera, Maui. One of about 20 of the remaining species of the family Amastridae, thought to have once included over 325 species. Photo by David R. Sischo.

Plate 14
Laminella sanguinea, Oʻahu. The beautiful shell of this rare species is usually hidden beneath
a layer of the snail's own excrement, for reasons unknown. Photo by David R. Sischo.

Plate 15

Philonesia sp., Oʻahu. Photo by David R. Sischo.

Plate 16

Partulina physa, Hawai'i. A group of snails estivating in a tree hollow, a not unusual site for this activity. Photo by David R. Sischo.

But there is another important factor at play here. Of the 36 Hawaiian species in this family that taxonomists currently recognize, almost all are already extinct. In fact, Nori and Ken's extensive field surveys have managed to locate only two of these species. Within the context of a "triage taxonomy," they explained to me that "working on an amazing group like the endodontids is a lower priority because they're mostly gone." And so, even if some dedicated soul does one day take on this work, and there does end up having once been 300, or even 100, additional Hawaiian species in this family, it will most likely be too late to do anything for them but to add them to the growing list of the extinct.

The tragic situation of the Hawaiian endodontids reminds us that there are unknown extinctions going on around us all the time. When you think about it, how could things be otherwise? There are an estimated 8 million species yet to be described, and there is no reason to think that these unknown species are being spared the impacts of our current period of mass extinction. In fact, most scientists who have spent time thinking about how to make sense of the level of extinction among unknown species have reached the conclusion that, if anything, they're likely to be disappearing more quickly than the species we do know about. This is because it is thought that unknown species will tend to have smaller geographic distributions (which is generally correlated with higher extinction risk), will probably be disproportionately located in biodiversity hotspots (which are generally places of high habitat loss), and will not be the subjects of any targeted conservation effort.[40]

The vast majority of these unknown losses are, without doubt, invertebrates. Some of them will be snails. In fact, Hawai'i's snails are probably better studied than most, with many more undescribed species to be found in other parts of the world. In the case of terrestrial snails, the best estimates leave a great deal of uncertainty. They tell us that roughly 29,000 species have been described around the world, with anywhere from 11,000 to 40,000 remaining to be identified.[41] Given that, as one recent study put it, terrestrial gastropods are "arguably one of the most imperiled groups of animals" on the planet, it stands to reason that many of these species will already be gone if and when they are eventually described.[42]

In the last decade or so, a number of more concrete examples have emerged of recent snail extinction of previously unknown species. Some of these species have been discovered through specimens in museum collections; others have been discovered as shells deposited in dunes and other sites. Either way, these shells have turned out to be all that was left of their species at the time that it was realized that they were, indeed, members of a unique species.

In one recent study, nine new species from the Gambier Islands of French Polynesia, all of the family Helicinidae, were discovered. All of them had been collected on Cooke's Mangarevan Expedition and sat, undescribed, within the Bishop Museum collection since. Sadly, follow-up surveys by researchers revealed that all of these new species, along with 33 of the islands' other 36 snail species, are now extinct, most likely as a result of deforestation and habitat loss.[43]

Similarly, on the island of Rurutu in French Polynesia, work over the last couple of decades has revealed an additional eight species of the family Endodontidae. Where there were once thought to be 11 species, we now know of 19. Here too, however, these species were able to be discovered only in shell form. The researchers concluded: "The radiation of endodontids in Rurutu was thus much larger than previously envisaged. However, we hypothesize that all species of the family are now extinct in the island."[44]

It is hard to make sense of these unknown extinctions, of species that are discovered already lost. A species bursts into existence, ready to be named, described, and hopefully appreciated; but at the same time, it is already a former species, the only record of its having lived at all, a lingering remnant like a shell. We can make intuitive sense of it when these discoveries involve fossils—brontosaurus and mammoths that roamed the earth long before our time—but it is somehow a more unsettling prospect when these losses become contemporary companions.

But, as odd as the phenomenon of unknown extinctions may seem, the numbers tell us that it is actually the norm—indeed, overwhelmingly so—with roughly four times as many undescribed species as there are described ones on the planet. I suspect that the contemporary prevalence of this mode of unknown extinction comes as something of a surprise to many people precisely because we are so fixated on the small portion of the animal kingdom that is well described.

We are living in the midst of an unknown extinction crisis, one that is stripping innumerable species out of the world before we have even realized they're here. The incredible loss of diverse plants and animals that we know about, that we can name, that we can make some partial sense of, is only one side of the coin. While there is without doubt much that remains unknown—even actively ignored—about all of the species being lost today, it is vital that we appreciate that there is something else going on here as well: an unknown extinction crisis that is at once both larger and more thoroughly beyond the scope of our comprehension.

What is unique, then, about stories of snails that went extinct before discovery is not this fact in itself, but rather that their existence and extinction have come to be known about at all. In most cases, when the last of a species dies, all record of it vanishes with these individuals. These are what the philosopher Michelle Bastian has described as "extinctions never known."[45] We might think, for example, of the diverse ecologies of soil biota that have almost certainly disappeared with changes in agriculture that have pumped more and more chemicals into our soils. Or the complex communities of invertebrate specialists that once fed on the whale carcasses that fell to the deep-sea floor, before commercial whaling led to their removal from the oceans. Within recorded history, countless species once found in these and numerous other ecosystems have no doubt disappeared without a trace.

But snails have a particular advantage over many other species when it comes to being discovered post-extinction. Unlike the majority of other invertebrates, whose soft bodies mean that they frequently leave no earthly trace after death, snails possess a remarkable calciferous remainder. In their shells they leave behind a record of their presence, even if a thoroughly imperfect and incomplete one.

A snail shell is a miraculous thing. For hundreds of millions of years snails have been wandering the planet—its oceans, rivers, and lands—their fleshy, porous bodies protected by these sturdy calcium carbonate structures. These shells provide a record of a life. The apex, or innermost point of a shell's spiral, is its oldest part. When a terrestrial snail is hatched or born, it begins life with this tiny shell. As it grows, it secretes calcium carbonate and other chemicals from its mantle, to build up around the aperture and incrementally extend the outermost whorl of the shell. Unlike arthropods and many other invertebrates that have an exoskeleton that must be shed to allow growth, snails have evolved protection that can expand with them, never requiring a period of vulnerable exposure. But this protection comes at a price: this shell consumes as much as half of the energy that snails require for their growth.[46] If you allow your eyes to trace the spiraling pattern of a snail's shell outward, from apex to aperture, you are retracing the life history of this tiny being.

In a variety of ways, these shells can be read to provide information about the lives they once contained, a condensed, solidified, and "embodied history."[47] The thickness of a snail's shell can vary greatly depending on the nutrition available in the environment. Periods of life in which growth ceased altogether can be marked with a little scar, called a "varex." At a much longer temporal scale, shells also record some of the features of the life of a species, with specific adaptations reflecting things like the habitat they occupied, their dietary specializations, and the predators they lived among. For those who are able to read them, shells can not only announce the former presence of a now extinct species; they can also provide us with some important glimpses into its former existence.

To be sure, other creatures can also leave important traces after extinction. With the aid of climate-controlled museums, even the tiniest and most fragile species are now sometimes able to be discovered long after they have disappeared, perhaps a butterfly pinned to a board or a sample of leaves and flowers pressed between pages. In this regard, the plants and animals that fascinated the collectors of the past—including Hawai'i's larger snails—are the ones whose unknown extinctions are most likely not to remain that way.

But, courtesy of their shells, snails are among that small club of invertebrates that don't require museums in this way. As Ira Richling and Philippe Bouchet, the biologists who discovered the nine new snail species in the Gambier Islands, have put it: "Documenting extinction when it has taken place even before scientific collecting is limited essentially to vertebrates, snails and, to some extent, crustaceans. These taxa have in common that they leave postmortem remains (bones, shells and carapaces) that can be traced in the archaeological record or in discrete soil or cave horizons."[48]

There are no precise figures available on how long snail shells endure in this way; a great deal depends on the size and thickness of the particular shell, as well as the soil, climatic, and other environmental conditions. Even up in the topsoil, things vary considerably. In some conditions, shells have been shown to be badly degraded in as little as a few months, in others they are thought to last decades, perhaps even a century.[49] Tucked away a little deeper, though, as subfossils, some land snail shells have been shown to survive largely intact for tens or even hundreds of thousands of years.[50] We will encounter some of these kinds of shells in the epilogue.

In this way it seems to me that snails—more than most other living beings, perhaps more than any other—have the capacity to interrupt the pervasive phenomena of unknown extinctions, to draw our attention toward and allow us to see previously unnoticed losses. As a highly diverse and highly threatened group with a durable remainder, snails find themselves positioned somewhere between the invertebrates and the vertebrates: possessing the abundance of species found among the former, and the hard architecture of many of the latter. It is this unique position that makes snails not only an emblem, but with a little luck potentially also a powerful disrupter, of our rapidly unfolding unknown extinction crisis.

CARE FOR THE UNKNOWN

Among all the snails that I encountered at the Bishop Museum, perhaps the most surprising were to be found in an old wine fridge, tucked away in a quiet corner of the Division of Malacology. They are surprising primarily because unlike the countless empty shells with which they share these rooms, these snails are alive and thriving. Their continued existence is the decades-long labor of the biologist Daniel Chung. In the late 1980s, Daniel noticed that many of the small, inconspicuous, ground-dwelling snail species were disappearing. At this point, the colorful *Achatinella* tree snails were just starting to gain some conservation interest. There certainly wasn't any government or popular concern or resources for these far less showy snails. Alongside being tiny in size, many of them are dark brown in color and generally covered in dirt and rotting vegetation anyway. In short, they're a particularly hard sell when it comes to limited conservation dollars.

So, Daniel decided to do something. As he explained to me when we met one afternoon at the museum, he probably should have applied for permission, but he decided that it was better to act swiftly and ask for forgiveness later if need be. And besides, he noted, "no one really cares what people do with invertebrates anyway."

Daniel and I chatted while he prepared new leaves for his tiny charges. All of the snails he looks after are detritivores, consumers of dead and decaying leaf matter. While it is hard to know for certain, many of them are thought to be exceedingly rare; some are probably even extinct in the wild. Roughly 20 species can be found here, most of them belonging to the family Amastridae. As previously noted, only 23 species in this endemic Hawaiian family are thought to remain, with over 300 having already been lost.

Daniel initially cared for these snails at home. He figured out how to keep them alive by trial and error; eventually, he learned how to help them thrive. At first, they lived in a modified refrigerator, a DIY version of the environmental chambers in use at the SEPP captive breeding lab. Later, he moved them to the wine fridge that is now their home. For about the last two decades, most of these snails have been at the museum, taking advantage of the site's relatively secure and stable climatic conditions.

Watching Daniel work that day, as he removed each small plastic container from the fridge, replaced its vegetation, and returned it to that lonely box in the corner of the room, it was hard not to reach the conclusion that these snails have been prematurely made museum specimens, relegated to the very margins of the conservation world by their status as particularly unremarkable invertebrates.

And yet, they are known, even if only by a few people. They are cared for and kept alive in the world, even if in this somewhat unconventional and informal way. In fact, with the growth of interest in conserving all of Hawaiʻi's snails in recent years, the significance of this little community living quietly in a corner of the museum has begun to be much more widely appreciated.

Take, for example, *Amastra intermedia*. These largish snails with their conical brown shells would probably have been extinct if it wasn't for Daniel. In 2015, the last known free-living snail—a single individual—was collected by Dave and the SEPP team. This little being might have ended up like George, the last snail of *Achatinella apexfulva*, living out its final days in solitary confinement in the captive breeding facility. Luckily, this was not to be. Daniel had collected two of these snails over a decade earlier. One of those snails died soon after, but the other—perhaps through selfing or stored sperm—produced offspring. From that snail, Daniel was able to keep the species going in captivity. In 2015, six snails from Daniel's population were combined with the one that the SEPP team had collected. They have now produced hundreds more, and enabled the establishment of populations housed at the museum, the Honolulu Zoo, the SEPP ark, and in one of the forest exclosures, located nearby to where that final wild snail was collected.

This is precisely the kind of conservation that can't happen without detailed taxonomic work. Among other things, it requires knowing the species, identifying their decline in time to pull them into captivity, and knowing enough about them (or their close relatives) to keep them alive in these conditions so that, one day, they can be released back into the same kind of habitat that they once occupied.

Amastra intermedia is but one in a long list of Hawaiian species that have benefited in recent years from a growing body of knowledge about the islands' snails. This work is vitally important, and yet we are also seeing that it is not enough. The situation described in this chapter makes clear that while targeted conservation efforts can help many species, for the foreseeable future there will continue to be innumerable others—just how many, we don't know—that will slip through the cracks of these approaches.

Our current planetary predicament requires a concerted effort to get to know as many of these creatures as possible, not only to document their existence but to understand their conservation status and their needs. This is a point that has been being made for decades now as interest and funding for taxonomy have declined in many parts of the world, leading some to speculate—with a healthy dose of hyperbole—about the extinction of taxonomy itself.[51]

But these efforts to know are also not enough. The reality is that in the coming decades innumerable undescribed species will become extinct—we simply will not be able to find and describe them, let alone produce meaningful knowledge about them, in time. As such, an adequate response to the unknown extinction crisis demands something much more radical than the cultivation of a new category of "charismatic microfauna," in which gastropods and insects might join pandas and elephants in an extended club of conservation poster children. To take invertebrates seriously is to acknowledge that there are simply too many living kinds on this planet for our practices of conservation to be tightly constrained by those we know in any meaningful way. Instead, we must cultivate our capacities of appreciation and care for the unknown.

For decades, some conservationists have criticized the focus of both public messaging and conservation spending on charismatic mammals and birds. In some cases, they have argued that other species, like ants and snails, ought to take the spotlight for a while. But for the most part they have instead insisted that we need to be thinking beyond species. From this perspective we should be focusing on conserving the ecosystems that provide habitat to these and countless other species, known and unknown.

In response, it has often been pointed out that endangered species conservation generally aims to do just this. Charismatic animals often function as "umbrella species," with efforts to protect their habitat providing shelter to myriad others. There is definitely something to this argument, notwithstanding the fact that not all charismatic species end up providing particularly good umbrellas for others; a lot depends on the overlap in distributions and specific needs of the species in question.[52] But to the extent that these umbrella species function as intended, the distinction between prioritizing species or ecosystems might not be a very significant one. It might, as Jean Christophe Vie, the deputy head of the IUCN's Species Programme, has put it, be a case of "doing the same thing but with different packaging."[53] And when it comes to compelling packaging, whales and elephants will likely win out every time.

This is only the case, however, if endangered species are conserved in situ, within their habitat. In the case of Hawai'i's snails, as will be discussed in the final chapter, there are significant—perhaps insurmountable—obstacles to conserving them in the forest or returning them to these places. Here, the combination of habitat loss and introduced predators has led to a situation in which "evacuation," as Dave has described it, has become the only option for survival. By its very nature, however, this kind of ex situ approach must conserve species one by one. It relies on identifying and extracting known species, and so unavoidably misses almost all species that are as yet unrecognized (the odd microscopic hitchhiker might benefit, like the feather lice species inadvertently conserved with some captive bird populations). At the same time, of course, this kind of conservation does very little for the many known plants, insects, birds, and other species—the broader ecosystem—of the Hawaiian forest.

Ultimately, it seems that in our current situation what is needed is a combination of species-centric and ecosystem-level approaches: conserving individual species in a targeted manner, preferably in situ, when they have become so threatened that this is now necessary for their survival, but at the same time, conserving whole ecosystems and landscapes.

Of course, including an ecosystem-level focus doesn't solve the problem of unknown species. In many ways, in fact, it demands even greater knowledge of the relationships and processes that must be considered and conserved. What's more, if we want to have any hope of knowing whether these conservation programs are actually working, we need a good sense of who inhabits these places and how our actions are affecting them. And so, while it is certainly possible to proceed with conservation in the absence of anything like a complete "Encyclopedia of Life"—indeed, we must—there is no getting away from the necessity of basic taxonomic research.

Returning to those drawers full of shells in the Bishop Museum, I am more convinced than ever both of the importance of getting to know this diversity intimately and of the need for modes of appreciation that exceed our knowing. Years later, I find myself not much better equipped as a taxonomist than I was at the outset of this journey, albeit with a much greater appreciation of the significance of some of the things that I do not know enough about, from shell morphology to taxonomic practice. While working to know more matters profoundly—it challenges us and expands our capacity to conserve and live well with others—in this era of escalating loss, our knowledge cannot mark the limits of our care. In addition, we need to find new, expansive ways to make sense of and to make connections with the vast and threatened worlds of all those incredible beings that are buzzing, crawling, whirring, and, yes, sliming around us unseen and unappreciated.

5 THE EXPLODED

Solidarity and Military Snails

I slowed down and pulled off the highway in front of a large chain-link fence. As I approached the open gate, a man in army camouflage signaled me to pull over. I lowered my window and he asked for my name. Checking it against the list on his clipboard, he waved me on into the small parking area of the Mākua Military Reservation.

It was a little before 7 am as I parked the car and got out. The first rays of sunlight were just making their way over the back of the valley. In front of me stretched a sea of grass surrounded by sheer walls of jagged rock, deeply etched by erosion. Mākua Valley, almost 5,000 acres in size, had the feel of a giant amphitheater, vast yet contained. The grasses and low bushes that covered the floor of the valley extended most of the way up its walls. Around the top rim, at least in patches, I could make out some areas of forest. It was among these trees that the last of the valley's larger tree snails once lived.

Over the next half hour, I took in Mākua Valley from the parking lot as other people slowly arrived. We had been told not to wander, so I kept with the group, occasionally chatting with others, eager to learn more about this place. When we had all arrived, Vince Dodge, who would be leading our cultural access visit, called us all together for introductions. About 15 people formed a loose circle. Vince welcomed us to Mākua, explaining that for generations this place had been home to Hawaiian people, making their lives within the valley's expansive embrace. Despite its relatively hot and dry climate on the leeward side of O'ahu, the valley floor had been extensively farmed. At least three significant heiau (temples) were constructed in the valley in ancient times, and the moʻolelo tell us that this whole area was considered to be wahi pana, a sacred place, one intimately associated with Hawaiian stories of creation. Mākua—which means "parents" in ʻōlelo Hawaiʻi—is thought by many to be "the site where *Papa* (the earth-mother) and *Wākea* (the sky-father) meet."[1]

Traditional life in the valley changed significantly after the overthrow of the monarchy. From the early 1900s, with the completion of a railroad, much of the land in the area was acquired by the McCandless Ranch, in many cases after their free-roaming cattle had made the area unviable for kuleana (small land parcel) farmers.[2] In the 1940s, however, things took another major turn as both the ranch and any remaining farmers were summarily evicted. In the aftermath of the bombing of Pearl Harbor, during the turbulence of World War II, Mākua and the surrounding areas were appropriated by the US Army under the authority provided by a declaration of martial law (Hawaiʻi was at this stage a US territory, rather than a state).[3]

The valley was enrolled into the war effort, to be used for live-fire training exercises, including everything from mortar and artillery fire to the detonation of bombs and even bombardment by missiles fired from air and sea. Over the subsequent decades, Mākua also became a dumping ground for the Army, with a large section at the back of the valley serving as an "open burn, open detonation" site where the military disposed of everything from conventional explosives to mustard gas, napalm bombs, and large quantities of white phosphorous.[4]

This sacred place carries the scars of this destructive activity. Some of these scars are dramatically visible in the landscape, such as the large strips of denuded earth and invasive grasses that now stretch themselves up its walls. As a result of decades of explosions and their subsequent fires, the complex plant communities that once lined this difficult terrain have been lost. While some of this destruction had been well and truly set in train before the Army arrived, especially as a result of cattle ranching, decades of live-fire training have significantly degraded what was left.

But many of the other scars of this activity are harder to see. They take the form of toxins that have leached into the soil and the water in ways that are impossible to really quantify, especially as most of the relevant information is not publicly available. They are also there in the vast quantities of unexploded ordnance that litter this place, lying hidden within the head-height invasive grasses that have taken over the damaged landscape, or buried just below the surface of the soil, waiting to be triggered.

As a result of these legacies, spending time in Mākua Valley is now an inherently hazardous activity. And yet, twice a month for roughly the past 15 years, the community group Mālama Mākua has run cultural access visits here. These visits last most of the day, taking anyone who is interested and physically able to a variety of cultural sites dotted around the valley. In our group that day were environmentalists, peace activists, academics and students, and a number of local residents who just wanted to know more about this strange fenced area that they had driven past on the highway so often. A few members of our group were Kānaka Maoli, but most of us were not. These cultural access visits, I discovered, are as much about education as they are about caring for and building an ongoing relationship with this place.

This is, without doubt, an odd arrangement. The US Army is not in the business of granting cultural access to active military facilities; in fact, I have been unable to identify any other such instances. And so, it is perhaps unsurprising that in Mākua this arrangement was not really entered into voluntarily. Rather, it is the somewhat unhappy and unstable product of decades of activism and litigation by Mālama Mākua and others, opposing the Army's occupation and destruction of this place. Alongside cultural access, these negotiations have also seen a complete cessation of live-fire training in the valley, although the area does continue to be used for other, less explosive, military purposes.

Given these hazardous legacies, it is not surprising that trips into the valley come with strict conditions. After our group had finished introducing ourselves, one of the five Army personnel standing off to the side joined our circle for a mandatory safety briefing. Just in case the release of liability forms we had been required to sign to enter this place had left us with any uncertainty, we were strongly reminded that this is a dangerous site, riddled with surface and subsurface explosive devices.

Our instructions were clear: we were to stay in a group, always ensuring that we walked between the Army vehicles that would be at each end of our convoy. An explosive ordnance disposal (EOD) specialist would be up front, scanning the terrain. We were also to stay on the dirt road at all times, unless directed otherwise; in fact, we should stay toward the middle of the road because heavy rain tends to wash explosives into the gullies that form along the verges.

At a few selected and preapproved sites, we were told, we would be venturing off the road. In those places, we must stay between the yellow ropes. The roped-off areas had been recently checked for hazards. In addition, however, some of the places we would be visiting in the valley were deemed by the Army to be fragile archaeological sites. As such, staying within the ropes was also about protecting the valley from us, from the impacts of our boots and our hands. To this end, we were told that our group would also be accompanied by two Army archaeologists whose job it was to ensure that these sites were not disturbed.

At this suggestion, I noticed a couple of the other members of our group roll their eyes. I can only assume that they, like me, were wondering about the Army's moral authority to position itself as protector of a place that they had spent the better part of the last 60 years blowing up.

This chapter explores the recent history of Mākua Valley. Unsurprisingly, it does so with a particular focus on the now-missing gastropods. Since the early 1980s, in particular, there has been an ongoing struggle over the biological and cultural heritages of this place. The knowledges, relationships, activist strategies, and technologies that have emerged through this process have had an enormous impact on the conservation and future of snails throughout Hawai'i and beyond.

The situation in Mākua also provides a concrete way into a much more pervasive environmental issue: militarism. All over the United States, and indeed around the world, the US military has established bases and training facilities that impact profoundly on the environment. When it comes to endangered species, the track record of the Department of Defense (DoD) is questionable at best: clearing their habitat, destroying their nesting sites, or just blowing them up. These impacts are so substantial, in part, simply because the DoD controls so much land. Within the United States, roughly 25 million acres are managed by the military, making it one of the largest landholders in the country.[5]

Fascinatingly, these military lands also turn out to be particularly rich in endangered species. In fact, when compared with other federal lands, including national parks, DoD lands have been shown to have both the greatest number and the highest density of species listed as threatened or endangered under the Endangered Species Act (ESA).[6] Across the country, roughly 400 such species are found on DoD lands, including 15 species of snails. While at least a few of these species got to be endangered in no small part as a result of living on military lands, for many other species these places, despite their many challenges, now offer a better chance for survival than the available alternatives.

As a result of this strange situation, the US military now often finds itself thrown into the role of conserving endangered species. As a federal agency, subject to the ESA, the various branches of the DoD are required under Section 7 of that act to ensure that "any action authorized, funded, or carried out is not likely to jeopardize the continued existence of any endangered or threatened species."[7] In and around Mākua, satisfying these obligations has required the Army to develop an extensive program of snail conservation, including the building of most of Oʻahu's snail exclosures, alongside endangered plant nurseries and out-planting programs.

But there are numerous similar examples of Army conservation programs. We might think of the streaked horned larks and the Mazama pocket gophers that make their homes at the Joint Base Lewis–McChord in Washington State. Or the tiny Saint Francis' satyr butterflies and the red-cockaded woodpeckers of North Carolina's Fort Bragg, thought to be the world's most populated military base. In more or less comprehensive ways, military operations at these and many other sites have been required to accommodate rare plants and animals, and the DoD has had to foot at least part of the bill for their active conservation.

It is very easy to tell simple stories about the US military as either an environmental villain or a savior. Indeed, both of these kinds of stories are frequently told in the media. The reality, however, is more difficult. The snails of Mākua offer an example of some of the ambivalent and multifaceted relationships between conservation and militarism.

Hawaiʻi is a particularly important place to take up these topics. In this archipelago, as we have seen, conservation is even more underfunded than in much of the rest of the United States, with nearly one-third of the country's listed endangered species found here, but less than 10% of the allocated federal funding for their conservation.[8] At the same time, parts of these islands are among the most heavily militarized locations on the planet: Oʻahu alone is home to seven major military bases and around 50,000 active-duty personnel. And so, it is perhaps not surprising that the DoD's own assessment has found that the four military facilities across the entire nation with the highest number of threatened and endangered species are all found in Hawaiʻi; in fact, all are on Oʻahu. Mākua is number two. A total of 168 at-risk species are found in these four locations. This situation is a testament to the incredible endemism of Hawaiian biodiversity. But it is also a result of long legacies of environmental destruction in these islands, coupled as they are with ongoing militarization on a staggering scale.

In Hawaiʻi, as we see in Mākua Valley, this situation is further complicated by the fact that it plays out on colonized land. Here, struggles over the future of snails and bombs are inextricably caught up with the efforts of Kānaka Maoli to secure rights to land and culture, to imagine and enact sovereignty in diverse ways.[9] In some cases, as has been documented in these islands and around the world, the needs and understandings of Indigenous peoples come into conflict with those of conservationists working to protect endangered species. Sometimes these differences can be resolved; in other cases, they are seemingly intractable.[10]

The story of Mākua Valley, however, offers an example of a powerful alignment of goals that has brought together a diverse constituency of Kānaka Maoli and other local community members, conservationists, lawyers, and activists working for demilitarization as well as social and environmental justice. Importantly, in Mākua, these people have been brought into solidarity not only with one another but with and through the valley's remarkable snails.

While my aim in telling this story is to move beyond a simplistic, black and white, account of the Army's impacts on snails and other endangered species, I must acknowledge from the outset that I take it as given that a sustainable future for these islands, and for the planet more broadly, is fundamentally incompatible with the current size and operations of the US military. My focus in this chapter is on military managed lands, and even in this respect alone the situation is stark. In addition to the huge areas of land occupied within the country's borders, the US military has a vast network of roughly 800 bases on foreign soil (by way of contrast, Britain, France, and Russia have a combined total of about 30 foreign bases).[11] Maintaining this global footprint, alongside all its other activities, has made the US military the largest institutional emitter of greenhouse gases in the world, so much so that it single-handedly produces significantly more emissions than many medium-sized countries like Sweden and Aotearoa/New Zealand.[12]

And yet, in a valley on this little island in the middle of the Pacific Ocean, this war machine has been brought to a standstill now for over a decade by a coalition of people in solidarity with a snail. I tell this story because there are important lessons to be learned here, but also because it offers hope that something better than the status quo might still be possible.

THE SEARCH FOR SNAILS

With the safety briefing completed we set off from the parking lot, making our way north across the mouth of the valley. Our first stop was at an ahu, an altar, crafted out of black lava stones. Here, we left offerings people had brought for the purpose—water, flowers, leaves, and songs—and we announced our intention to travel into the valley. Despite its weathered look, this ahu is of more recent construction than one might assume. Not being allowed to interact with the ancient archaeological sites in the valley, and not wanting to disturb them anyway, in 2001 Mālama Mākua negotiated with the Army for permission to build three ahu as focal sites for the community and visitors, to acknowledge and show their respect for this place.

When everyone had left their offerings and said their piece, we continued on our way. Turning onto a dirt road that runs along the northernmost edge of the reserve, we began slowly and steadily climbing toward the back of the valley. To our left, the valley wall rose up above us, covered in grass with a thin band of trees along its upper edge. But to the right we could see all the way across Kahanahaiki and Mākua Valleys. Although the area is widely referred to as Mākua, it is actually made up of three valleys, with the two aforementioned valleys being the bulk of the area. From many angles, the division that separates these valleys is not immediately obvious, but from our position that morning on the northern edge of Kahanahaiki, the ridge that juts out between them was unmistakable.

Seeing this ridge plainly for the first time, I remembered reading a report from the early 1980s that noted that it was once home to "a particularly rich population" of *Achatinella mustelina*.[13] This report was written in the years immediately following the listing of all of the snails of the genus *Achatinella* under the Endangered Species Act in 1981. Historical surveys by the biologists d'Alté A. Welch and others showed that *Achatinella mustelina* was once found in this area.[14] So, the Army somewhat reluctantly called in the biologists to see if they were still here. The Bishop Museum was given the contract, and Carl Christensen, who was then in charge of the museum's malacology collection, took on the job with Mike Hadfield, assisted by Peter C. Galloway and Barbara Shank.

It was Mike who first told me about Mākua Valley, early on in my research. He described to me his time spent searching for snails in the difficult and unusual conditions afforded by this place. Even in the 1980s, the remaining forested pockets of the valley were confined to its edges, gulches, and sheer sides, as well as up around its rim. This is difficult terrain at the best of times, but in this case, traveling off the well-worn paths, they were pretty likely to encounter unexploded ordnance. The team underwent basic training to identify these hazards: snail scientists learning taxonomic skills for shells of a very different kind.

Their task was made all the more challenging—even if safer—by the fact that they also had to be accompanied by EOD specialists. This meant that, as Mike explained to me, every time one of the team wanted to look in a particular tree, they'd have to announce their intention and then wait for the Army personnel to take a look first. This made the whole process incredibly time consuming. But it certainly wasn't an unnecessary precaution. Over the course of their survey work they encountered many signs of the Army's long history of activities, including an unexploded rocket and a 1,000-pound bomb, likely one that had been sitting in the landscape since World War II.

But there, in that most unlikely of places, they also discovered the snails they were looking for. At higher elevations, especially above 500 meters, they encountered numerous *Achatinella mustelina* living quietly among the trees. In some places around the rim of the valley they were still present in very high numbers. The largest population of all, however, was found at the back of the Kahanahaiki–Mākua division ridge. Mike and Carl noted in their report: "the authors know of no locations where Oahu Tree Snails occur more abundantly."[15]

While it was only this one species that they had been sent into the valley to survey for, as it was the only federally listed species known to exist in the area, the team also recorded the presence of a variety of other snails. Among them were living members of the genera *Philonesia*, *Tornatellides*, and *Succinea*, as well as shells—and so perhaps also reclusive individuals—of the species *Amastra rubens* and *Auriculella ambusta*. Many of these species are themselves exceedingly rare, albeit not formally listed.

Even more surprising, however, was an encounter with a single individual of the species *Partulina dubia*, a large tree snail. While archaeological evidence shows that several species of this genus once lived on Oʻahu, all of the others are long gone. When I asked him about this discovery, Mike recalled the scene: "I heard Carl exclaim, 'son of a bitch.' I looked over and saw him staring closely at a hole in a tree trunk, repeating to himself, 'son of a bitch, son of a bitch.'" Having reviewed the historical collectors' records in the Bishop Museum prior to going into the field, Carl had been on the lookout for the species, but he told me that he hadn't really expected to find it. The last time the species had been sighted was before he was born, over four decades earlier. Sadly, their chance encounter with this single individual was to be the last time the species has been seen. *Partulina dubia* is almost certainly now extinct.

Alongside this hidden wealth of snails, the survey also revealed the extensive impact of the Army's activities. In a few locations, the team encountered groups of *Achatinella mustelina* living in trees that now had shards from exploded rockets embedded in their branches and trunks. Some areas of forest had clearly been blown apart, and with them their rare and endangered snails.

But more significant than the explosions themselves were the fires to which they gave rise. The team encountered several areas of burned forest that would formerly have been home to snails, including evidence of fires to "a considerable elevation" up the Kahanahaiki–Mākua division.[16] Equally as concerning as this direct loss of habitat, Mike and Carl began to worry that these fires might be pushing rats and other snail predators up into higher elevations, a fear that was reinforced by the large numbers of empty, broken *mustelina* shells they found on the forest floor in these areas.

Their report made a variety of recommendations to the Army; chief among them was the need to limit and better control fires, as well as to modify the designated "high explosive impact area" so that it did not include actual or probable snail habitat. In the years that followed, however, not a lot changed. As the respected environmental journalist Patricia Tummons summed it up: "the fires showed no letup in either frequency or size."[17] But armed with reports like this one, Mike and other conservationists continued to lobby the USFWS to enter into a formal consultation with the Army under Section 7 of the ESA. While some of the other endangered species in the valley figured in these discussions too—a variety of rare plants and a single bird species, the Oʻahu ʻelepaio—it was the tree snails that became something of a poster child for the struggle over Mākua.

Around this time, Mike also established a field site on the rim up above Mākua Valley in the Pahole Natural Area Reserve. Here, he and his colleagues and students began to conduct extensive mark and recapture studies of *Achatinella mustelina*. It was quickly apparent that they were monitoring the decline of the species, primarily as a result of predation by rats and wolfsnails. From this vantage point, Mike was also constantly reminded of the destruction unfolding in and around the valley. In the late 1980s, he took Michael Sherwood, a lawyer from the newly established Hawaiʻi office of the Sierra Club Legal Defense Fund, up to the site. Sherwood began to add legal pressure to the Army and the USFWS, both to do the research to develop a fuller understanding of environmental impacts and to work to mitigate them.

In one particularly strange alignment of circumstances, it was Carl Christensen who drafted the initial letter to the Army that was signed and sent by Sherwood. Having given up on a career in malacology for the time being after he was let go by the Bishop Museum, Carl had enrolled in law school at Harvard University. Four years later, in 1989, he was back in Hawai'i over the summer vacation, interning with the Sierra Club Legal Defense Fund. "One of my tasks," he told me, "was drafting a letter for the attorney to send to the Army saying that they should have listened to what those guys Hadfield and Christensen had told them years before about burning up the valley."

But the Army hadn't really listened, and in the years that followed it continued not to. There were some limited changes as a result of this legal pressure in the late 1980s and early 1990s: for a time the Army halted some of its more destructive activities—like the helicopter gunnery practice that was said to be starting many of the fires—and made some efforts to improve its firefighting and containment practices.[18] In the mid-1990s, however, something about the nature of this pressure shifted, and the cause of the snails began to align with the activities of a group of people living at the other end of the valley, on Mākua Beach.

MOBILIZING SNAILS AT THE BEACH

We made our way steadily upward, along the northern edge of Kahanahaiki. Although it was still early, the heat in the day was already building and it quickly became hard work. As we climbed higher, the view opened out. I stopped to catch a momentary breeze, and turned to look back down the dirt road. From this elevated position, I could see across the bottom of the valley, over the highway to the beach and the ocean beyond.

In the past, the area we were now in was divided into two distinct ahupua'a, a traditional Hawaiian land parcel that often runs all the way from the mountains down to the sea, ensuring that the occupants of each piece of land had access to a diverse range of resources, from mountain plants to ocean fish. Today, the Farrington Highway skirts the Wai'anae coast, cutting through these traditional divisions. All of the extensive lands mauka (on the mountain side) of the highway are controlled by the Army, while the little area makai (on the sea side) of the highway, only a few meters wide in places, remains a public, or at least semi-public, beach.

It was by this beach, a few days earlier, that I had met with Uncle Sparky Rodrigues. Our meeting had been arranged by Justin Hill after I had made contact with Mālama Mākua through their website, wanting to talk to them about their work and its connection to the snails. Justin had replied, inviting me along on the cultural access visit and offering to arrange a meeting with Uncle Sparky. The three of us met by the side of the highway, next to the little Mākua Beach parking area. After some brief introductions we made our way along a path into a small area overflowing with trees and other plantings. Uncle Sparky pointed out some old garden beds and mosaics, remnants of the dedicated labor and care of his late wife, Aunty Leandra Wai.

As we walked that day it had begun to rain, and at Uncle Sparky's suggestion the three of us headed under the thick canopy of a kou tree. Uncle Sparky began to tell us stories of this place. I learned that the struggle over the valley did not end at the Army's fence. Over the past several decades, the beach area has also been a site of intense contestation. Locals have used this beach for fishing and recreation, but it has also been a place where people, especially Kānaka Maoli, have built their own homes and found peace and protection. As far back as the 1960s, however, numerous efforts have been made to demolish these homes and evict residents to make way for everything from film crews and amphibious Army training to proposed state parks.

Despite obstacles, this community persisted across decades, rebuilding each time their homes were demolished.[19] Uncle Sparky explained that all kinds of people lived at the beach in the past: "elders, people that were sick, people that were crazy, couples, singles . . . Society called them homeless. They were not homeless; this was their home." Many of these people, after having been priced out of housing in the surrounding area, turned to Mākua as a last resort. But the place became something much more than this.

Aunty Leandra and Uncle Sparky lived at Mākua in the 1990s. He explained to me that "Mākua was our place to heal." They were having difficulties in their marriage at the time. When they first came to Mākua, like many of the other residents, Aunty Leandra started cleaning up. She removed decades of accumulated rubbish, much of it dumped by businesses and private individuals, and she worked to push back the introduced grasses and plants like koa haole that had taken over. Alongside their houses, these new residents planted food crops like sweet potatoes, pumpkin, and beans, as well as a variety of other native and traditional plants.

"The ahupuaʻa we're in is called Kahanahaiki," Uncle Sparky told me, "and it is a place of transition, and that transition is where you can come to heal." In the past, he went on to explain, a kumu (teacher) or kupuna (elder) might have helped with that healing process, whether physical, mental or spiritual. "Here, we didn't have that . . . so the ʻāina is the one that helped us. . . . Over the years, Mākua helped us heal our family, heal our relationship, to the point where we started becoming more aware of what was going on around us."

In the mid-1990s there were about 300 people living at Mākua Beach. Again, there was growing pressure to evict residents, and Uncle Sparky and Aunty Leandra became leaders in the efforts to resist. As they organized, they also became increasingly engaged in broader struggles over the Army's occupation of the valley and the damage that it had inflicted on the ʻāina. Their concern was about damage to cultural sites as well as impacts on the forest and the wider environment within the valley. But, crucially, it was also about the potentially toxic impacts of these activities—especially the open burn, open detonation practices—on the coastal lands and waters that they lived on, fished in, and otherwise relied on. "We started finding out about the poisons, the residue of all the military activity, poisoning the land."

It was also at this time that they learned about the snails. A group of professors and students from the university had come to the Waiʻanae area to an information event where Uncle Sparky, Aunty Leandra, and others were speaking about the struggle for Mākua. "One of them," Uncle Sparky told me, "was Hadfield." Mike told them about his work on the snails and his efforts to get the live-fire training in the valley stopped. That fortuitous encounter sent the residents of Mākua to the Sierra Club Legal Defense Fund where, Uncle Sparky told me, "we asked them about the snails." In the months and years that followed, they learned about the declining prospects of *Achatinella mustelina*, how restricted their home ranges are, how long it takes them to reach sexual maturity, how many predators they face in the forest. "So," Uncle Sparky explained, "they were at risk and under tremendous threat. They became part of our mantra as well: How do we protect the native snail?"

In their subsequent activism over the beach and the valley, a genuine interest in the well-being of these struggling snails was accompanied by a strategic decision to mobilize the legal power afforded to endangered species under US law. After years of trying to find legal representation to push the case of the Mākua Beach community in the courts, they had concluded that "we can't go there to defend us, but we can go there to defend endangered species." Mālama Mākua was born in 1996, bringing together concerns over Kanaka Maoli rights to land and culture, over the militarization of the islands, and over the health of the environment. In short, as the name implies—mālama meaning to care or protect—the group came to represent a coalition of people and interests with a shared focus on protecting this place and its future.

Mālama Mākua took up the slow, difficult work of challenging the Army's activities in the valley through the courts. In this effort they have been supported by David Henkin, an attorney from the Sierra Club Legal Defense Fund (now Earthjustice). Most of these legal actions have been brought under the National Environmental Policy Act (NEPA) and have centered on the need for the Army to determine the scale and significance of their impacts in the valley by conducting a detailed Environmental Impact Statement (EIS). In pushing their case, Mālama Mākua has emphasized impacts on the snails and other endangered species, but also the host of other cultural, social, and health consequences that are potentially associated with the destruction of this place and its lingering toxic legacies.

A QUESTION OF EXPERTISE

Our little convoy made its way along a narrow, overgrown path in the rear part of Kahanahaiki Valley. We were now traveling single-file, between a set of yellow ropes, having left the Army vehicles that were traveling with us behind on the main track. On all sides we were surrounded by tall grass, well above head height. At a few spots the path went down into a small gully, and we had to scramble up and down through the mud. Eventually, we came to a heavily treed area and all of us, I think, were thankful for the shade.

We were making our way toward an important cultural site, referred to as the piko stone. As our group slowly arrived, we gathered together in the shade. We sat and rested, taking in the place. Vince spoke to us about the site. It had become a significant place for him in part because it had been one of Aunty Leandra's favorite places in the valley. Nearing the end of his remarks, Vince explained to the group that he wanted to leave hoʻokupu (offerings) here. Doing so, however, would require him to step outside the yellow ropes, to get closer to the stones. At this point one of the Army archaeologists traveling with us reiterated the rules: we must stay between the ropes to ensure that the archaeological site is not impacted on by our movements.

Vince, clearly frustrated but not surprised by this response, calmly turned to us and told us that this was a valuable opportunity to understand the difficulties of this working relationship with the Army. At the heart of his frustration was a fundamentally different orientation to the valley and these sites: where the Army sees relics of the past to be preserved as they are, Mālama Mākua sees a living heritage that should be respected and cared for through ongoing connection and acknowledgment.

At the same time, Vince explained, the sheer hypocrisy of the Army's position was hard to stomach: "This is the Army's concept of protecting these sites. They bombed the shit out of them for 60 years, and now we're coming to reconnect, and we're being prevented." But in his view these restrictions were also not stable and consistent. The Army's pronouncements around the wearing of shoes, the leaving of offerings, the sites that are open and closed, are constantly shifting and generally not open to negotiation. "This is what we push up against constantly," Vince explained.

This exchange highlighted an ongoing tension over understandings and expertise in the valley. Over the past several decades, the Army has brought more and more archaeologists and biologists in-house. When I asked Kyle Kajihiro about this situation later, he summed things up succinctly: "It is about controlling the expertise; owning the means of knowledge production, to dispute both what Hawaiians and environmentalists might say." Kyle is a long-term member of Mālama Mākua and an activist and academic whose work focuses on demilitarization.

In his writing on this topic, Kyle has argued that at Mākua the Army learned from the Navy's struggles over the island of Kahoʻolawe. For over four decades, the US Navy used this island just to the southwest of Maui for target practice, raining bombs and missiles down on it from the sea and the sky. The ultimately successful struggle to halt these activities in the 1970s— including a series of occupations of the island by Protect Kahoʻolawe ʻOhana (PKO)—has become an iconic chapter in Hawaiian history.[20] In Kyle's view, where the Navy was caught off-guard to some extent, the Army deployed a more sophisticated program of information and population control in Mākua. After all, he explained, the Army has "more experience with counter-insurgency." In this war on endangered species and sacred sites, controlling information and public perception is everything.

On that particular day, Vince decided that it was best to abide by the Army's decision, at least for now. Stepping over the rope, we were told by the Army personnel, would likely have resulted in the cancellation of the cultural access and all of us being escorted out of the valley. Sometimes, Vince said, we do need to push these issues to maintain rights in the face of ongoing changes and challenges. In determining when and how this is appropriate, he follows the guidance of the valley. And so, Vince set his hoʻokupu down on a rock just outside the roped area, without leaving its confines, and the altercation came to a close.

Just a few kilometers east of where we gathered that day, another crucial event in the recent conservation history of Hawaiʻi's snails took place. It, too, was centrally concerned with the establishment of the expertise and resources required to control the future of this valley. Up above us, directly adjoining the military reserve, lies the state government–managed Pahole Natural Area Reserve. As noted in chapter 1, it was here that Hawaiʻi's first snail exclosure was constructed in 1998. The exclosure was built by the state government in the area around the rim of Mākua Valley that had been the focus of Mike's *Achatinella mustelina* research since the early 1980s. In fact, this first exclosure literally encircled within it the 5 × 5 meter mark and recapture quadrant within which he had been monitoring the slow-motion destruction of the population, at that point reduced to fewer than 20 snails.

It was around this same time that the Army began to shift gears in its conservation efforts. A small team of scientists was established by the Army in 1995 to better monitor impacts on snails and other endangered species across the island. Mike agreed to include some of these scientists on the USFWS permit that allowed him to study the federally listed snail species. The Army's scientists began to conduct their own monitoring of snail populations on and around Army facilities, including Mākua and on the other side of the island adjacent to the Army's Schofield Barracks East Range. At the same time, they worked to actively manage threats to these snail populations, including through fencing and removing ungulates that destroy the forest, and establishing predator control programs for rats.

As part of this effort, inspired by the Pahole exclosure—and using some leftover materials from it—the Army built one of its own at Kahanahaiki. This new construction was less than a kilometer from the state's Pahole exclosure, just across the border into the Mākua Military Reserve. Through these activities the Army shifted from simply limiting the impacts of its own activities on endangered snails—or at least trying to, in some very limited ways—to becoming an active manager of their populations.

One of the key pressures for this change was the Army's entering into formal consultation with the USFWS over Mākua Valley. While this process technically commenced in 1998, in the years leading up to it the pressure had been growing and there had been several periods in which the Army had been called on to account for its actions. Various conservationists had been pushing for a formal consultation for some time, as had Mālama Mākua. In fact, earlier this same year, David Henkin, acting on behalf of Mālama Mākua, sent a letter to the Army notifying them of their intention to sue if such a consultation was not entered into.[21] Very shortly after that letter was sent, training exercises in the valley sparked a massive 800-acre fire—one that burned up onto the ridgeline that divides Mākua from Kahanahāiki.

This was the final straw. David explained: "Seeing the writing on the wall, the Army immediately stopped all military training at Mākua Military Reservation and 'voluntarily' entered consultation with the US Fish and Wildlife Service."

This consultation quickly reached the conclusion that Army operations in Mākua were impacting a range of threatened plant species, as well as *Achatinella mustelina*, and that these species should have a targeted Implementation Plan developed to protect them. To this end, in 1999 the Army created an Implementation Team composed of their own scientists as well as representatives from various government, university, and conservation organizations, including Mike. This team would be responsible for ongoing research and consultation to help the Army identify additional measures that it might take to stabilize these species "with sufficient numbers of populations to ensure their long-term viability."[22]

While there were at the time 10 surviving species of the genus *Achatinella* listed under the ESA, the Army's responsibility extended only to those species directly impacted by its activities. As only *Achatinella mustelina* is found in this area, it was the only snail considered in these negotiations. Importantly, the Army's responsibility is only to "stabilize" this species, not to "recover" it. In order to satisfy this requirement, the species need not be secure and self-sustaining; instead, it just needs to no longer be declining toward extinction.

But this consultation with the USFWS also required that the Army conserve as much of the genetic diversity of the species as possible. To this end, in 2000 genetic studies were commenced drawing on samples from snails in 18 locations around the Wai'anae Range. These studies were primarily conducted by Mike and Brenden Holland—in fact, it was this project that brought Brenden to Hawai'i as a postdoctoral researcher. From this work it was determined that the species is composed of eight "evolutionarily significant units" (ESUs). As discussed in chapter 2, snails tend to be infrequent dispersers and so once isolated often do not to get back into contact. As a result, each of these ESUs is composed of a population or group of populations that is, in effect, already drifting off in a different evolutionary direction. In fact, when I asked Nori about *Achatinella mustelina*, she told me that she wouldn't be surprised if future taxonomic analysis reveals that it is actually already several distinct species. Even without this information, however, the Army was required to commit to conserving each of these distinct ESUs. Much more than a single exclosure would be required to do the job.

At the same time that conservationist pressure was building on the Army, things also began to come to a head with Mālama Mākua. Finally, after years of delays by the Army in conducting an EIS, in July 2001 the federal court in Honolulu issued a preliminary injunction that prohibited the Army from carrying out any training in the valley while the case was pending. This was a very significant outcome. As Mālama Mākua's attorney, David Henkin, explained to me: "To my knowledge this was the first time a court had ever ordered the US military not to do training because of a violation of an environmental law."

It was only a few months after this that the attacks of September 11, 2001, in New York and Washington took place. The Army was adamant that they needed to get back to training in the valley and entered into negotiations with Mālama Mākua to make this possible. By the following month a settlement had been reached. The Army could resume very limited live-fire training each year for 3 years, but, David explained, "if they got to the end of the third year and hadn't done the EIS they would have to stop training until they finished it."

In 2004, the 3 years had elapsed, and the EIS was still not completed. As a result, all live-fire training in the valley was halted. In the almost two decades since, little has changed on this front. To this day the EIS has not been completed to the satisfaction of the courts and live-fire exercises in the valley have not resumed.

Alongside this end to live-fire training, the settlement reached with the Army in 2001 also made a range of other concessions to Mālama Mākua, concessions that did not have a sunset clause and so are still in place. The Army agreed to start some cleanup in the valley, to conduct further studies, and to provide a technical assistance fund for the local community to hire outside experts to peer-review these studies. Cultural access visits, like the ones that I have joined, were also a key part of this settlement. Twice a month, Mālama Mākua takes groups of members, local residents, and visitors into the valley in this way. Twice a year, there are also overnight visits in the valley for the Makahiki festivals, organized by a broader community group called Hui Mālama o Mākua.

Vince described these cultural access visits as "the game changer, the toe in the door." He explained: "We knew cultural access was really important because of what happened on the island of Kahoʻolawe. As long as Kahoʻolawe was unreachable, it was lost. But once there was cultural access and people started to go . . . it totally changed the balance of power. Because people realized, they knew experientially, that this place is alive, and we need to help take care of it."

In the early years of this struggle, Mālama Mākua was led by Aunty Leandra, Uncle Sparky, and Uncle Fred Dodge (Vince's father). Fred was a local physician and one of the early opponents to the Army's activities in the valley, alongside being a passionate advocate on a range of other community issues in the Waiʻanae area. Although not of Kanaka Maoli descent, as Vince explained it to me, his father "had a calling from Mākua from the first moment he set eyes on her in 1962."

It was Aunty Leandra and Uncle Fred who, for about the first decade, led the cultural access visits to the valley; one full day, twice a month, every month. Today, as the original kupuna who steered this project have passed away or begun to find the strenuous cultural access visits too much, a new generation have taken up the mantle. Vince has stepped in to take on some of the work that his father did. Aunty Lynette Cruz, a professor of anthropology and longtime advocate for Hawaiian cultural history and rights, has taken on the role of president of Mālama Mākua. At the same time, the group have initiated a kiaʻi (guardian) program in which young people are helping out and being trained to lead cultural access visits. In this way, Mālama Mākua is laying the foundations to ensure its ability to continue to care for the valley into the future.

Throughout all of Mālama Mākua's work, the snails have played a vital, albeit often largely unseen, role. In particular, David Henkin explained to me, they were one of the key grounds on which the court relied in 2001 in issuing the preliminary injunction. Unlike many of the other environmental impacts of the Army's operations—which were difficult to prove, let alone quantify—beginning with Mike and Carl's survey in the early 1980s, there was firm evidence that *Achatinella mustelina* was present in the valley and that Army training was destroying their habitat. As such, there was an indisputable need for a full EIS to be developed. While the Army tried to argue that they were not likely to push endangered species over the edge of extinction, this was not accepted by the court. As David explained: "the court correctly responded that this is not the standard of significant harm; you don't need to be threatening to wipe things off the face of the earth to have significant harm."

All these years later it remains somewhat unclear why the Army has not completed the EIS and pushed to resume live-fire training. Some people I spoke to thought that Mākua had simply become too difficult: the growing local awareness and opposition that has been fostered by Mālama Mākua and others—including of endangered species and cultural sites—has made it more of a hassle than it is worth. At the same time, the Army's priorities in Hawai'i and around the region have shifted. They have needed to establish other training areas and to monitor their impacts in other places, all of which have absorbed considerable time, energy, and resources. With this in mind, other people felt that Mākua had simply slipped to the bottom of the list, and as such that the struggle may flare up again at any moment.

Whatever the reason, Mālama Mākua has welcomed the years of peace in the valley. But the group knows that even if live-fire training never resumes, life in the valley may not ever return to a demilitarized state. The Army's lease on Mākua ends in 2029, and there is a growing insistence within the community that the time has come for this land to be returned. Even if this were to happen, though, the legacies of decades of abuse will continue to haunt this place in the form of unexploded ordnance, toxins in the soil and water, as well as, of course, the many missing species. It may ultimately prove to be impossible to return the valley to a safe enough state for public access to really be possible.

Shortly after the Army took control of the valley in 1941, it signed an agreement with the territorial government that it would vacate the area 6 months after the conclusion of the war, and that when it did it would "remove all its property and return the premises . . . [in] a condition satisfactory to the Commissioner of Public Lands."[23] Over the decades since, the Army has consistently reneged on this agreement to leave, despite pressure from the territorial and then state governments. At the same time, it has gradually worked to minimize its obligation to restore the area. At present, under the terms of its current lease with the state, the Army is required to restore the valley only "to the extent that a technical and economic capability exists and provided that expenditures for removal of shells will not exceed the fair market value of the land." Furthermore, the Army has been discharged of all liability and claims relating to this restoration work.[24]

Mālama Mākua see their next big task coming in the struggle to hold the military to account on this restoration work. Their goal, as Uncle Sparky expressed it, is to "get them to clean it up so it's not another Kahoʻolawe"—a site that, while it has technically been returned to the state of Hawaiʻi, is not considered to be safe for human inhabitation. In fact, at the time that the Navy completed its cleanup work, 25% of the island had not been remediated in any way and was deemed to be unsafe and off-limits, while the vast majority of the rest of the island had received only a surface treatment. Ongoing cleaning is now being undertaken by the state with the dedicated labor of volunteers.[25]

The tactic deployed by the Navy at Kahoʻolawe is one that has been widely used by the DoD, both within the United States and all over the world. It even has a name: military to wildlife conversion. Rather than exhaustively restoring an area, it is simply converted into a wildlife reserve or other conservation area. Often these lands are handed over to the USFWS; in the case of Kahoʻolawe, they were given back to the state to become the Kahoʻolawe Island Reserve. When former military facilities are converted in this way, the remediation standards are considerably lower, and therefore cheaper to meet. If people will not live in the area permanently, levels of toxins and other hazards can be higher; some problematic areas can simply be fenced off. Alongside the economic advantages, the military gets all the positive PR associated with establishing a new environmental reserve. A recent survey identified more than 20 former US military bases that have been redesignated in this way, taking in more than a million acres of land, and this includes only the federally managed lands.[26]

Mākua Valley, with its high numbers of endangered species—even if many of them are only barely surviving—might make a good candidate for this kind of conservation area. While such a conversion might ultimately be the best option, Mālama Mākua wants to see the valley cleaned up properly before any such decisions are made. While conservationists around the United States now often embrace these spaces and their possibilities, this enthusiasm frequently stems from the lack of alternatives within "ever-diminishing habitat reserves."[27] In this scenario, conservation receives the scraps, damaged scraps at that, that fall from the table of the war machine.

It is also important to note that these processes of military to wildlife land conversion are not part of a general downscaling or reduction in US military land use. Rather, they are one of the gradual retirement of lands that become too damaged, difficult, or controversial, in favor of fresh lands to be despoiled. In fact, a 2014 study found that "the U.S. military has been increasing its training estate by approximately 1,200 ha per year."[28] In this way, ever more land is pulled into this process, made unviable for human and nonhuman life, littered with problematic legacies for generations to come, pulled in at a rate of one new Mākua Valley roughly every 18 months.

Mālama Mākua hope that they have done enough to draw the attention of the islands, and the world, to this little valley in the middle of the Pacific Ocean. They hope that they have done enough to ensure that, when the time comes, the military is not able to simply walk away.

While the future of Mākua Valley remains far from certain, one thing is clear: the protracted struggle over this place has had significant benefits for the snails. The lawsuits, settlements, and public scrutiny have not only halted most destructive activities; they have also pushed the Army into a more active conservation role, albeit sometimes with some odd consequences.

Since the mid-1990s, the team of Army scientists working on the conservation of snails and other threatened species on this island has steadily grown. From a handful of people, it is today a team of over 50 scientists and technicians called the Oʻahu Army Natural Resources Program (OANRP). This group is responsible for the Army's compliance with a range of federal conservation regulations across its various training facilities on the island. Mākua Valley, as the location of the most contentious and highly visible struggle with the Army in Hawaiʻi over conservation, was central to the growth of the OANRP and the Army's reluctant acceptance of these responsibilities.

For roughly the past 25 years, the OANRP's snail work has been headed up by Vince Costello, a passionate and committed advocate for all of Hawaiʻi's diverse gastropod species. The central component of those efforts has been the construction of a network of exclosures on Oʻahu, beginning with the one at Kahanahaiki. With each new exclosure the OANRP team has refined the design and the barriers, borrowing ideas from organic gardening approaches and even the escargot industry.[29] The three barriers in operation at the Palikea exclosure, which I described in chapter 1, are now standard: the angle barrier, the cut-mesh barrier, and the electric barrier. Systems for real-time monitoring are also constantly being refined to notify staff if, for example, a branch falls across the exclosure fence, allowing possible access by predators.

As a result of all this work, now stretching back over more than 20 years, the US Army has become one of the most significant funders of snail conservation in Hawaiʻi. It isn't possible to say just how much is spent on this work. In 2018, however, the Army reported an expenditure in that year alone of roughly $5.8 million on the Mākua Implementation Plan (of which the snails are a significant part). With similar spending each year for about two decades now, and an ongoing commitment into the foreseeable future, the costs are racking up.

One of the most bizarre elements of this work is that, at least insofar as snails are concerned, all of this funding is technically directed at a single species. While numerous other snail species rush headlong toward extinction, each of the eight distinctive genetic populations of *Achatinella mustelina* is having an exclosure built to conserve it.

Even the other federally listed species receive nothing like this level of funding. Despite the fact that seven of the nine remaining critically endangered *Achatinella* species are found in the Koʻolau Range on the other side of the island, until very recently there was only a single exclosure located in those mountains (today there are two). As the Army has a much smaller impact there, they have not been required to undertake mitigation work.[30] Similarly, the four islands of Maui Nui—where, again, the Army has not been required to mitigate its impacts in the same way—have an equivalent number of snail species, including three listed under the ESA, but until the last couple of years have had only a single exclosure.[31]

My point, then, is a simple one: while exclosures are a vital technology for saving disappearing snails, it is the Army that has had the resources available to build the bulk of them, and this situation has seen them distributed in a way that is shaped more by the vagaries of Army operations than it is by the needs of snails. In fact, more specifically, it is shaped largely by the particular history of activism and legal challenges to the Army's operations on behalf of the snails.

It is hard to know what to make of all this. Certainly, the OANRP biologists that I have met, who are carrying out this work on behalf of the Army, are passionate about the future of snails. They are doing the very best that they can in deploying the available resources to save species. This is a view shared by some at Mālama Mākua. As Uncle Sparky put it to me: "it's the only department that I truly love about the military. I'll tell people to get out there and volunteer with them because they're doing really good work." For the Army as an organization, however, it is pretty clear that this conservation work is simply a means to an end: ensuring that training operations are disturbed as little as possible. At the same time, the Army takes every opportunity to publicize the environmental projects that they have so often been forced, through years of costly external pressure, to undertake.

Ultimately, however, the reality is that these snails would be disappearing with or without the Army's impacts. Wolfsnails would still be up in the forests of Mākua Valley eating them out of the world, as they are all over the Hawaiian Islands. The sad and entirely illogical result of this situation is that your best chance of survival as an endangered snail in Hawai'i is to be a member of a species that is being, or has been, routinely blown up by the US military.

Importantly, the Army's conservation efforts have ended up benefiting many other snail species. The exclosures that have been built around the needs of *Achatinella mustelina* have in practice become bubbles of safety for the other species that happen to also reside in these areas. The OANRP and SEPP teams actively coordinate to get as many of them as possible into these spaces. But the fact remains that their placement has been designed almost entirely around the specific needs of only one species.

While the Army is almost certainly having an overall positive impact on the snails, this is far from being an optimal arrangement. More fundamentally, we might ask why endangered snails should have to rely on military funding at all? Why is this the only part of the US government with these kinds of funds available to it? Surely, in any rational arrangement of resources it would make sense for conservation agencies and research institutions—who are the principal holders of the relevant expertise, and the organizations for whom this is a primary responsibility, not a means to an end—to be funded adequately to take on this work themselves. This, however, would require the Army to relinquish precisely the control over expertise that it has worked so hard to craft.

SNAIL SOLIDARITIES

Returning to the main track from the piko stone site that afternoon, we began to slowly make our way back downhill, toward the front of the valley. The next time I visited the valley, about a year later, I walked this same stretch of track with Kyle Kajihiro. Grounded in his long-term work with Mālama Mākua and a range of other organizations focused on demilitarization, as well as his academic research on the topic, Kyle explained to me the need to understand the dispute over this particular valley in a much larger context. As he put it:

You can't talk about militarization in Hawai'i without its broader connections. The reach of American militarism is global. Hawai'i is the hub of Pacific Command, so anything that is happening here is affecting other islands and other places in the region. And what happens there has repercussions back to Hawai'i.

Mālama Mākua has worked to develop an approach that acknowledges this interconnectedness, promoting efforts to share information and build networks of solidarity. As Army and other military operations are effectively blocked in one location, they inevitably move elsewhere. As Kyle pointed out to me, victory in the struggle to halt the Navy's destruction of Kaho'olawe meant that some of this activity was pushed to other sites. The same thing has happened with Mākua:

> The fact that we've stopped the live-fire in Mākua means that Pōhakuloa [on the island of Hawai'i] is absorbing a lot of this activity. So, we're actively making connections with the folks there and trying to build their capacity to fight it. Even if it might slow down progress with our site, with Mākua. I think it's an obligation we have.

Other members of Mālama Mākua that I spoke with shared this view, referencing connections with related struggles at Kahuku on the other side of O'ahu, at Maunakea on Hawai'i, and of course at the military's Pōhakuloa Training Area, where bombing continues to this day. Some of these sites now appear on the Mālama Mākua website under "connected struggles." As Aunty Lynette, the president of Mālama Mākua, explained this work to me, it is about "standing in solidarity, not just against the military, but in support of social and environmental justice."

Our group made its way to the final stop on our cultural access visit. Now almost back at our starting point by the highway, we took a sharp turn to the south, skirting an area of well-mown grass to make our way down into a small gully. There, under a canopy of trees, we were brought face to face with a large petroglyph rock (ki'i pōhaku), standing about 3 meters high. Human figures danced across the rough gray surface, alongside other forms that I could not quite make out, all of them carved long ago by the residents of this place.

While the whole valley is a place of learning, over the past few years this particular site has become a kind of informal classroom. All sorts of people, including musicians, researchers, and peace activists, have been invited to come along on cultural access visits to share their insights in the presence of the petroglyphs and the valley. Often these classes help to make connections between this place and the broader issues surrounding militarism. In addition to reaching the people on the cultural access, including both Army personnel and other visitors, this learning also makes its way out of the valley through Mālama Mākua's website and videos posted to YouTube.

It was via just such a video that I was able to hear a short talk by Ann Wright that took place at this spot. Wright is a retired US Army Colonel and diplomat who, in 2003, resigned in opposition to the Iraq War. On that day she told visitors to Mākua about the roughly 800 US military bases dotted all over the world, and about the trauma that they cause for local people and the environment. Bases in the Pacific, the Middle East, and Europe have all been sites of local conflict. Wright spoke about one such struggle, currently underway, over efforts to build yet another runway for the US military on the tiny Japanese island of Okinawa, into the habitat of endangered dugongs. As pressure has mounted over the US presence in Japan, the military has increasingly looked to Guam and the Northern Mariana Islands instead. A massive troop redeployment is now underway. Part of this redeployment, as Wright noted, is a proposal by the Navy to use the largely uninhabited Pagan Island for target practice, turning it into yet another piece of scarred and sacrificed land, with all of the associated long-term legacies for local people and the environment.

Among the many other precious plants and animals that call the tiny island of Pagan home is the beautiful, critically endangered, tree snail *Partula gibba*. Fascinatingly, these snails too have become central actors in projects of inter-island solidarity for demilitarization. Over the past 10 years, Mike Hadfield and Dave Sischo have both been involved in efforts to protect their island home in the Northern Marianas. Having been drawn to this place by *Partula gibba* in 2010, to determine for the USFWS if this species was still present, they have used their knowledge of local biodiversity to help oppose efforts to destroy the island. In addition to the US Navy's ongoing maneuverings, this has included another proposal to turn Pagan into a dump for debris from the 2011 Japanese tsunami.

David Henkin from Earthjustice has also become involved in this struggle, working with some of the many passionate local people who are opposing the destruction of their lands. Before he could take on this role, he had to talk to Mālama Mākua and the Chamoru activists in the Mariana Islands about the potential conflict between their cases. He discussed with them the sad reality that success in blocking military activity in either location might increase the pressure on the other. "I'm pleased to say," he told me, "that both client groups acknowledged that and felt that everyone deserves good legal representation; they are in solidarity with each other." This legal avenue is open in the Mariana Islands because they are a US commonwealth and therefore subject to federal laws like NEPA and the ESA. This is not the case for the majority of the US military's overseas bases, where local people and species are left to rely on the ability of their governments to push back against the world's largest, and best funded, military.

As I headed out of the valley, I turned to take one last look at the incredible landscape that is Mākua. Within those rugged walls is a place of immense beauty and destruction, a place shaped by a deep and ongoing history of dispossession and violence, but also a place of resurgence and resistance. The relationship between militarization and conservation that is playing out here is a complex one, and one that is inseparable from other struggles over land, over culture, over the future of these islands and our planet.

While they continue to endure in the world, snails and other endangered species can play vital roles in efforts to protect land and culture from the ongoing and escalating ravages of militarism, resource extraction, overdevelopment, and more. But they don't and can't do this on their own. As the long and difficult history of Mākua shows, the agency and influence of snails is worked out in solidarity with others: with Kānaka Maoli, conservationists, activists, lawyers, and more, drawing on the resources of legislation and litigation, of community mobilization, of careful snail surveys and long-term population studies. Among all their other significances, then, to lose Hawai'i's snails is to lose a vital ally in the ongoing struggle for a more livable world.

Personally, my hope is that one day these efforts might reduce the size and scope of the world's militaries, and so their impacts on diverse lives and landscapes. Until this happens, however, these struggles play an important role in holding the military in the United States and elsewhere to account for their actions, ensuring that they go some way toward meeting their obligations to conserve species, to restore damaged lands, to enable the continuity of Indigenous cultures, and more. After all, these are obligations that the military's inflated budget prevents other agencies and local communities from having the funding to themselves address.

In its own way, Mākua Valley has joined this global struggle for demilitarization. The expertise, networks, knowledges, and passions formed in this place—many of them formed around and with the valley's disappearing gastropods—are reaching out into the world to help protect other endangered species, other people, other places. Strange but powerful snail solidarities are emerging.

6 THE CAPTIVES
Hope in a Time of Loss

The four of us stood gathered around the workbench counting snails, counting *Achatinella lila*, to be precise. The adults were easy enough to spot among the vegetation spread out in front of us. The keiki, however, were a different matter altogether. Their tiny forms, not more than a couple of millimeters in length, could easily be missed, mistaken for a mark on a leaf or twig. For this reason, every piece of vegetation had to be checked twice, by two different people. In all, there were 40 snails in this particular container, 18 adults and 22 keiki. As we finalized the count, the snails now gathered into two Petri dishes, I took the opportunity to get a closer look at their beautiful brown-green shells, glistening under the laboratory lights.

I was with Dave Sischo again, along with two of his colleagues—Lindsay Renshaw and Kimber Troumbley—this time, at the captive breeding lab of the Snail Extinction Prevention Program (SEPP). Alongside the network of forest exclosures, this lab is the other core part of SEPP's ongoing effort to hold open space for snails in the islands. While the last known free-living population of *Achatinella lila* disappeared from the Koʻolau Range a few years ago, the species has been able to live on here. An assortment of other rare snail species also make their homes in the lab, most of them either extinct in the wild or quickly heading that way. Drawn into the architectures of conservation care, these snails join a growing number of species around the world that are now only able to survive in captivity, and only with a great deal of assistance and care from people.

Adult snails of the species *Achatinella lila*, waiting for their cage to be cleaned.

This facility goes by a variety of names, more and less formal. Often it is simply referred to as the lab, but at other times I've heard it called the snail jail, or even the love shack. I've preferred to think of it as an ark—or perhaps more appropriately still, a kīpuka—a safe space, set aside from some of the worst of the destruction. A space that might, with a bit of luck, safeguard these life forms through this difficult time—and with them also hold onto at least some of their relationships and cultural associations—to allow this constellation of lives and meanings to one day be rewoven into the world beyond.[1]

This chapter is an exploration of the particular forms of care and hope taking form in this place. Keeping snails in the world in this way is a complicated process, involving dedicated and ongoing work. It is this work that has enabled the lab to become a site of hope: a space animated by the possibility that if the snails can just be sheltered through the storm, a brighter future might one day arrive. But there are no guarantees. As we have seen, for the vast majority of Hawai'i's remaining snail species, things are looking pretty dire. Even those snails tucked away in the relative safety of the lab remain at risk in a range of old and new ways.

This situation is compounded by the fact that it isn't clear how the lab's tiny charges might ever be released. At this stage, conservationists cannot even imagine the concrete steps that might be taken to make life in the wider forest possible for them again. In particular, all of the scientists I spoke to agreed that any such plans would require the widespread eradication of wolfsnails. But, as we will see, no one quite knows how this might be achieved. And so, the best that can be done is to hold onto rare snails in case something changes.

What becomes of hope in difficult times like these? This chapter explores the particular form of mournful hope that accompanies efforts to hold open room for snails in the midst of ongoing loss. In our Anthropocene epoch, a time of escalating environmental destruction, when hope is still available to us at all, this is the form that it must increasingly take. As the geographer Lesley Head has written of our current moment: "grief is a companion that will increasingly be with us. It is not something we can deal with and move on from, but rather something we must acknowledge and hold."[2] As we will see, in the work being done with Hawai'i's snails, hope is not a bold utopic proposition; rather, it takes the form of a refusal to give up care and responsibility in the face of escalating and inevitable destruction, and to keep working toward the best relationships and possibilities that are still available to us and others.

THE CARE OF CAPTIVES

The snail lab is a peculiar environment. Located on the eastern, windward side of O'ahu, in the Maunawili area, it is housed within a 13-meter-long trailer on a lush hillside just off the side of the highway. It is, to put it simply, not the most salubrious of environments. Dave, who had been good enough to give me a lift on the morning of my first visit, summed the site up nicely as we pulled up out front: "here we are, at our trailer in the swamp."

While the lab, like SEPP more generally, is run on an exceptionally modest budget, it is still a significant improvement on the past situation.

Prior to the establishment of this facility in late 2016, the only captive space for snails was the one that I mentioned in the introduction, where I first encountered George (a passion project set up by Mike Hadfield at the University of Hawaiʻi at Mānoa in the mid- to late 1980s, and staffed mostly by students and volunteers). As we have seen, over the past couple of decades, investment in snail conservation has slowly grown in the islands, initially with a particular focus on the larger tree snails that have official listing under the ESA, and today with a more inclusive approach to conserving all of the threatened snail species that can still be found (where available resources allow). This lab is a key part of these efforts.

Climbing a few metal stairs, Dave and I entered the trailer into a small office with a couple of desks, a coffee maker, and other essentials. The bulk of the interior space, however, was reserved for a large laboratory area. In contrast to its exterior, this space had a distinctly hospital-like feel: everything in its place, clean and sterilized, with a faint chemical smell hanging in the air. Here, small groups of snails live out their lives in plastic containers like the terraria you might keep a pet rat or fish in at home. Each of these containers, referred to as a "cage," is stocked with the appropriate vegetation for its particular inhabitants, be they leaf-cleaning, microbe-loving tree snails or detritivores. The cages themselves are stored within seven large environmental chambers that look quite a lot like fancy refrigerators but do the more complex job of replicating the specific temperature and moisture conditions required by each species. In this way, the lab aims to provide a series of simulated forest microcosms packed into one secure unit.

This place is a home away from home to over 5,000 snails of 40 different species. Keeping all of the lab's slimy inhabitants healthy is a detailed, time-consuming work of care. The bulk of this work falls to Lindsay, the lab's manager. Every 2 weeks, each cage needs to be taken out, censused, and refreshed. The first part of this process involves counting all of the snail occupants, which can take anywhere from 10 minutes to more than an hour. Cages with subadult tree snails are the easiest. These relatively large snails can be readily spotted, and as they are not yet breeding, there are no keiki to watch out for. Cages with breeding snails, like the one home to *Achatinella lila* that I helped census, need quite a bit more time and care, requiring that

An *Achatinella mustelina* cage at the SEPP Snail Lab.

everything be checked twice. Sometimes, a single keiki is missing, requiring 20 minutes or more of searching back through all the old vegetation until it is located.

But the most difficult cages are those that are home to snail species such as *Leptachatina vitreola* and *Cookeconcha hystricella*, which lay masses of tiny eggs. Censusing them literally requires the use of a microscope and tweezers to locate and collect the eggs that these snails carefully wedge into pieces of bark. Lindsay and the team keep close records, including snail and egg numbers and conditions. This way they can track and compare mortality and reproduction rates, and hopefully preempt any decline, or at least respond quickly if something starts to go wrong.

Once all of the snails in a cage are accounted for, it can be washed and steam sterilized. As Lindsay explained to me: "we're very big on sterilization here." Staff wear surgical gloves that are frequently changed, alongside liberal applications of ethanol on work surfaces, to prevent the accidental transmission of diseases or pathogens from one cage to another. After clean-ing, cages need to be rebuilt with the appropriate vegetation. For the tree snails, plant cuttings are collected out in the field every couple of days, often while on routine trips to the exclosures or other sites, and then stored in the fridge for later use. They need to be collected from higher-elevation forests to reduce the risk of them being host to pathogens spread by garden snails. But this vegetation can't just be thrown into the cages in any old way. There's a science, or perhaps an art, to it. As Lindsay put it:

> It's like floral arrangement; to build the right cage so that things aren't col-lapsing in on themselves. You've got to have the right airflow through the cage. Otherwise, things get mushy, things start decaying before they should, the leaves drop down to the bottom, and that affects the snails, they can start dying off from things like that. So, we've got to build it up just right.

After the vegetation is in place, any supplemental food is added. This food is prepared fresh every few days by the team: a potato dextrose agar medium is poured into Petri dishes and inoculated with a snail favorite, a *Cladospo-rium* fungus. Once refreshed, cages can be returned to the environmental chambers. But here, too, constant tweaking is required. Moisture levels can't simply be set and forgotten about. The unique combination of vegetation in each cage alters its moisture content, meaning that it can easily dry out or get too wet without monitoring. With about 70 cages in the facility, it takes 2 weeks to get through cleaning them all. When the last one is finished it is time to begin all over again. Keeping snails alive in this kīpuka is a detailed and dedicated labor of care.

Although underwritten by intense processes of loss, the lab represents an important site for the production and maintenance of hope: it safeguards within it the possibility that at some time in the future, after the wreckage has cleared, life might still be possible for these species. Hope is often thought about as a synonym for optimism, a desire for a particular state of affairs to come about. In reality, though, hope is a much more fickle and mysterious beast. As the writer Rebecca Solnit has put it: "Hope is an embrace of the unknown and the unknowable, an alternative to the certainty of both optimists and pessimists."[3] If we are certain that a desired possible future will arrive, or certain that it will not, then either way it cannot really be said to be hoped for. Hope relies on that space of ambiguity, of possibility, that resides between these two poles. In this way, hope calls on us to get to work, to do what we can to bring about a particular situation. In the lab, hope takes the form of the daily acts of attention and care that sustain living beings, and through them their species. It is an intensely grounded and practical hope, a practice of care for the future.[4] But it is not one without its risks and dangers.

As Lindsay closed up the refreshed *Achatinella lila* cage, its inhabitants all counted and safely returned, she gave the lid a few quick sprays of water. This seemingly insignificant act, she explained, has turned out to be one of vital importance. When cages are put back into the environmental chambers, and at the end of each workday, they get a light misting. This practice grew out of a speculation. Lindsay noticed that some keiki were sealing up on the underside of the cage lids and dying there. She also realized that the mist nozzles from the environmental chambers that go into the top of each cage don't always wet the lids themselves. So, she hypothesized that either these young snails may not be able to break out of their seals without an external water source or they might require that water as a cue to do so.

This indeed seems to be what was happening. Keiki were sealing up in their shells and dying of dehydration. Melissa Price, a biologist at the University of Hawai'i at Mānoa, explained to me that in the forest this sealing-up behavior would be beneficial, protecting small and vulnerable snails from drying out. In that environment they could rely on pretty regular rains coming through so they could unstick and crawl around and get some moisture. "It would be adaptive in the wild; it's nonadaptive in the lab." After this new procedure was put in place, keiki survival rates doubled.

More recently, another new protocol has seen a distinct decline in snail mortality at the lab, this time among adults. Previously, the leading cause of death for adults of some of the larger tree snail species was an unusual kind of cannibalism. At the heart of this situation is the humble mineral calcium.

It seems that some snails were developing calcium deficiencies, which led to blemishes on their shells. Other snails, also deficient in calcium, took this opportunity to scrape away at these blemishes to gain access to the calcium stored in the shell. Holes and wounds developed, leading to death. As soon as the situation was understood, the team developed an ingenious and equally strange response: they sterilized the shells of dead wolfsnails and introduced them to the cages to be consumed (after filling them with glue to ensure that keiki didn't get stuck inside them). In this way, Dave explained, "we're giving them back the calcium robbed from their ancestors."

Through the ongoing development and refinement of these kinds of protocols for care, most of the snails being raised in this new lab are doing well, at least most of the time. But, despite all efforts to better understand and respond to their needs, every now and then something goes wrong. One particularly significant incident occurred in September 2018, when a pathogen, parasite, or toxicant made its way into the lab. Lindsay explained to me: "we can't control everything. . . . We're bringing in outside vegetation, so sometimes things are going to come in on that." In this case, this seems to be precisely what happened. The snails in five separate cages, fed from the same bag of leaves, were affected. Individual snails became lethargic, their tentacles drooping, and they began to move around during the day when they should have been resting. Then they began dying in large numbers. Dave recalled the scene: "they died with their bodies outside their shells, as though they were trying to run from something." The team acted quickly, quarantining surviving snails in separate spaces—small plastic cups—with a piece of supplemental food. Lindsay and Dave held their breath to see what would happen next. Thankfully, this approach did the trick. Although many snails died, no species were lost. Things eventually stabilized and the remaining snails could be returned to sterilized cages.

A few years earlier, in a similar incident, most of the keiki of one of the detritivore species in the lab mysteriously began to die. Dave told me about the urgent search for answers: "We were rapidly trying to find out what was going on and finally we observed them under the microscope." There, mostly gathered around the snails' apertures, were tiny mites. Each snail had to be bathed several times under the microscope using a pipette to ensure that all the mites were removed. It was thought that the mites were likely introduced on the dead leaf matter eaten by the snails. As a result of this incident, the team now freeze and then thaw all vegetation for the detritivores before offering it to them; and all the snails get regular mite checkups under the microscope.

These ongoing events make clear that the lab is not quite the secure unit it might at first appear to be. Routines of care, record keeping, and quarantine have minimized pathogenic risks, but they can't eliminate them. In addition, there are a whole host of other potential problems to worry about. What happens, for example, if one of the environmental chambers breaks down? At similar snail facilities in other parts of the world, electrical malfunctions have led to significant mortalities.[5] Again, while some of these uncertainties can be moderated, they cannot be overcome. The facility has backup generators and an independent alarm system that monitors each chamber. If any of them goes 1 degree above or below their set points, all staff start receiving automated phone calls, text messages, and emails (through a cellular system that doesn't rely on the internet or the electrical grid to operate).

But what happens, for example, if the facility needs to be evacuated? The team has emergency plans in place for storms and fires; so far, they've worked, but they've also been lucky. In 2018, as Hurricane Lane approached the islands, all of the snails were relocated to the State Building in downtown Honolulu. Everyone was moved to safety in time, but it was a huge effort. When I spoke to Dave about it in early 2020, he was relieved but worried for the future. As he explained: "we only had 1,200 snails then, now we have over 5,000." A few months later, the reasonableness of that concern was confirmed, as Hurricane Douglas approached the islands. It seems, however, that where there is a will there is a way. Dave and the team managed, in a truly massive logistical operation, to relocate this larger population to the safety of Honolulu and then keep them cool and moist outside their environmental chambers.

In short, then, everything that can reasonably and affordably be done to ensure the security of the lab and its residents is being done. And yet all of this preparedness is also a reminder of just how easy it would be for this all to go horribly wrong, and just how high the stakes are. While this forest-in-a-trailer can offer reliable defense against wolfsnails and the many other key threats that Hawai'i's snails face today, in doing so it cannot help but create a host of new vulnerabilities. To put it simply, the lab is not a feasible long-term survival strategy. Despite all the labors of Lindsay, Dave, and the team, this bubble of care and the hopes that it sustains are intensely fragile.[6]

HOPE IN A TIME OF LOSS

No one wants the lab to be a long-term project. As Mike put it in our discussion of his work to establish the initial captive snail facility at the University of Hawai'i: "it was never my intention to create the world's largest snail zoo." And yet that facility and its successor have now been in continuous operation for more than three decades. Numerous generations of snails of diverse species have lived out their lives within these walls. What's more, the sad reality is that for most of the species in the lab—as well as those in the exclosures—life will not be possible outside these protected spaces in the foreseeable future. In fact, we cannot at this stage even imagine the steps that might be taken to enable release and recovery.

In almost all cases, returning species to the forest will require the eradication, or at least widespread removal, of wolfsnails. Of course, wolfsnails are only one of the relevant predators, but there is hope among some conservationists that at least some of Hawai'i's many threatened snail species may be able to cope with predation by rats and chameleons if the slime trail–following wolfsnails could just be taken out of the equation. This view wasn't shared by all, but it was unanimous that without the widespread removal of wolfsnails, it would be simply pointless to release other snails back into the forest; we'd just be feeding the predators. Removing wolfsnails from just one area, even a large area, would be only a temporary fix; in time they would simply reproduce and move back in. As Brenden Holland summed the situation up: "You'd have to do an island-wide purge . . . and I don't see a practical way of doing that at this point."

And yet Brenden, Dave, and other conservationists I spoke to were hopeful that in the future advances in predator control might improve the situation. Perhaps "hopeful" is not the right word. Perhaps it is more accurate to say that they wanted to be hopeful. While several possible scenarios floated around in these discussions, everyone agreed, without exception, that they were highly speculative prospects, ones that would not arrive any time soon, if at all, and that would likely come with their own considerable risks.

The most frequently mentioned possibility in these conversations was that of CRISPR and other genome-editing technologies. In recent years these approaches have received significant attention in all sorts of arenas, from possible new medical treatments for HIV and hereditary blindness to a slew of potential conservation applications. In Hawai'i, conservationists have also speculated about the possibility of controlling mosquitoes, and so the avian malaria that is decimating local birds, through this kind of intervention. While lab-based research along these lines on mosquitoes is now underway, there are considerable obstacles to development and eventual application.[7] In the case of Hawai'i's snails, however, no such work has even begun and the interest from researchers and funding bodies in controlling predatory snails, as compared to disease-spreading mosquitoes, is likely to be considerably smaller. Dave and others who raised this possibility with me know this. At best, he told me, he hopes genome-editing approaches might be available in 20 years or so.

When pressed on how things might one day get better for the snails, Brenden told me that he hoped that a pheromone-based bait might be developed that would allow a more effective and targeted approach to the eradication of wolfsnails, or at least make it possible to remove and control the species within some large sections of the forest in an ongoing way. At this stage, however, he told me that this prospect is nothing more than wishful thinking.

Meanwhile, other people have begun to champion the potential of a parasitic nematode that is widely used in agricultural pest control to kill a variety of snail and slug species. If released into the islands, they claim, it might kill wolfsnails. At this stage, however, there has been no research to determine whether this nematode would be effective against these particular snails, or whether it would in fact target the native snail species as well. Not surprisingly, given that it was a poorly planned and researched biocontrol initiative that got us into this mess in the first place, most of the biologists I spoke to in Hawaiʻi are approaching this prospect with extreme caution.[8]

Tucked away in their predator-free, climate-controlled lab, the snails are doing relatively well today. But we seem now to have arrived at a point at which most of Hawaiʻi's snail species will go on living in these captive spaces for the foreseeable future or they will not go on at all. Within these walls, some of these species might slowly decline and then peter out, or perhaps they will be taken in one fell swoop by a mystery pathogen. With the passing of George and the species *Achatinella apexfulva* in 2019, this possibility was made dramatically manifest. Other species might limp along in captivity, year after year. Others still may thrive: *Achatinella lila* is one such "success story." From a mere six snails brought into captivity about two decades ago, there are now hundreds and hundreds of them. But despite these seeming differences in vitality, all are in an important sense in a very similar situation. The species that go on reproducing at high rates are simply able to better disguise it. They, like all the others, have been held onto while their world has been lost, lost in a way that is, as far as we have any power to control or predict, irrecoverable.

If this lab is indeed an ark of some sort, how are we to understand this complex situation? What is an ark if the passengers can never disembark? What is an ark if the seas never stop rising, an ark that continues to be swept along in a growing storm year after year, taking on ever more passengers?

At least since the biologist Michael Soulé's classic 1985 essay "What Is Conservation Biology?" we have been accustomed to thinking of conservation as a "crisis discipline." In this context, endangered species management is frequently likened to emergency room work, accompanied by its own forms of intensive care, intervention, and even triage. On the surface the lab seems like a paradigmatic example of this kind of conservation. But perhaps this is not the case. Intensive care implies that there is a meaningful prospect of recovery, that with some attention species might be patched up and sent on their way. What happens when recovery becomes a less and less realistic proposition? For at least some of the snail species that now call the lab home, there is a strong chance that there will be no release. Just how many species will share this fate is something that we cannot know. What happens, in times like these, when many captive spaces like the lab are becoming something more akin to a hospice than an emergency room?

Hawai'i's snails are far from being alone in this regard. Indeed, this is the situation in which a growing number of other animals and plants find themselves today. Around the world, many individuals and their species live out their final days under human care in strange environments like zoos and captive breeding facilities, from giant tortoises in the Galápagos Islands and white rhinos in Africa to the diverse forest birds and snails of Hawai'i. With so many species at risk of extinction, as the situation gets more dire bringing all or some of the remaining individuals into these (relatively) safe spaces becomes an appealing option. But for many of them there will be no release. Reviews of reintroduction programs generally show that the majority of them fail for a variety of reasons, including an inability to secure suitable release habitat.[9] In these cases, maintaining captive animals becomes less an act of conservation than one of slowly drawing out an extinction. Unable to halt the ongoing destruction of our time, the Anthropocene has become a period in which both the lives *and deaths* of other species are increasingly being shaped, more or less wisely and consciously, by the actions of (some) humans.

What does hope look like in times like these? What does it mean to continue to care for species in the face of ongoing, unrelenting processes of loss? Lesley Head reminds us that hope does not require optimism.[10] We do not need to feel or believe that something is likely to come about in order to hope for it. Rather, hope is a practice of the possible. The hopeful work of the lab is not that of a bold, utopic vision for a future in which the world has been set to rights. Such a future, at least for Hawai'i's snails, is no longer imaginable. Rather, it is a practice of getting by on a damaged planet, while holding onto the possibility that something better—even if not better than now, at least better than what might otherwise be—can still be crafted with dedicated care and attention. The anthropologist Anna Tsing has referred to this as a practice of "gardening in the ruins."[11] It is a work of cultivating beauty and possibility as best we can, a work that insists that loss, even ongoing loss, does not negate obligations of care and responsibility.

Through our many hours of conversations I have reached the conclusion that this is something like how Dave sees the work of SEPP. He does the work of hope; however, his optimism is severely constrained. But, as he would also quickly remind me, no matter how compromised or limited the possibilities are that they can hold onto in the lab and the exclosures, without these populations there would be no hope at all. Indeed, Dave is actively working to expand both of these spaces, to build new exclosures and increase the capacity of the lab to take on yet more snails. What else can he do? At least while they go on living, these snails hold open the space of the possible.

But in times of loss we must also remember, as the philosopher Chantal Mouffe reminds us, that "hope can be something that is played in many dangerous ways."[12] Among some quarters of the environmentalist community, a core goal seems to be advocating for hopefulness and "good news stories," while avoiding pessimism.[13] As Head has put it: "There is a deep cultural pressure in the West not to be 'a doom and gloom merchant.'"[14] In this context, what is hoped for often seems less important than the act of being hopeful, of encouraging others into a particular orientation toward the future. But vague and general hope is not necessarily helpful. Hope is never innocent; it is always a question of which hopes and at what cost.

Keeping snails alive in the lab and in exclosures in a way that might enable some sort of future for them can, perversely, also end up enabling the continuation of destructive practices in their former habitats. We see this on military lands like Mākua Valley. While, as we have seen, the role of the Army in the conservation of snails is a highly ambivalent one, it seems important to note that the fact that endangered snails and plants are "backed up" in the lab and in exclosures positioned outside Army lands is one key part of what has enabled the military to continue its destructive practices in sensitive places for so long.

At the same time, we see another facet of the potential violence of hope in the way in which so many visions of a livable future for Hawaiian snails rest on the eradication of wolfsnails, as well as the ongoing killing of untold numbers of rats, chameleons, and other snail predators. Whatever we might think about the relative importance of these creatures' lives and suffering, the fact is that in this case our hopes for some rest on the destruction of others.

None of this is to say that we ought to give up on Hawai'i's snails, that our hopes for their future ought not to be sustained and nurtured. Rather, my point is a simple one: careful, responsible forms of hope require ongoing attention and scrutiny, for what they both enable and disable.

A MOURNFUL HOPE

As Dave and I drove back toward Honolulu on the afternoon of my first visit to the lab, we talked about some of the species missing from its chambers. Heading from the windward side of the island, our journey took us west on the Likelike Highway. There is one particular spot on this road, just as you're approaching the tunnel that takes you through the Ko'olau Range, that always leaves me in awe. The mountains rise up steeply in front of you, a sheer wall of lush, tree-covered green, is punctuated only by a series of sharp vertical ravines that have been carved by millions of years of erosion. It was on these incredible mountainsides, just a little further to the north in the central Ko'olau Range, that one of the snail species that almost made it into the lab once had its home.

As we drove, Dave told me about the loss, rediscovery, and loss again of *Achatinella pupukanioe*. The species was rediscovered in 2015 on a stretch of hiking trail thought to be completely devoid of *Achatinella* tree snails. In fact, no one had seen any of these snails in this whole region of the island since the 1980s. But there they were, in a single large tree. About 10 were seen, but there were probably more, quietly living out their lives unnoticed. Dave explained to me that they were easily spotted at night, but by day they would disappear into the curled-up leaves of their host tree.

> You'd go there in the day and you wouldn't find any. But at night they would come out and crawl around. The times we'd been there they were only on this one tree. We'd searched all around but never saw them on anything else.

They seemed to be a tiny, relic population. Having never encountered the species before, Dave and the team took photos back to Nori at the museum to be certain of the species identification, and later collected a genetic sample for storage. When it was confirmed that they were indeed *Achatinella pupukanioe*, a species thought to be extinct, Dave faced a tough decision.

At the time the current lab was not yet up and running. The University of Hawai'i lab was still in the aftershocks of a period of significant, unexplained mortality. It did not seem like a safe place to bring the snails. So, Dave made the decision to leave them undisturbed for now. But when they returned in 2017 to bring some of the snails into the new lab, like so many other snail populations before them, they had vanished without a trace. Recounting the tragedy, Dave told me: "We literally chopped the tree down and looked through every leaf. There was nothing." They carefully searched all the surrounding trees too, but to no avail. Not even a shell survived; "they were probably swallowed whole by wolfsnails," he told me.

Grief is a constant companion in the lab. The hope that has been crafted here is a mournful one, one haunted by that which has been lost. It does not seem at all a stretch to suggest that Dave, Lindsay, and many of the other passionate snail conservationists I've spoken with are living daily with a kind of "ecological grief"—a situation that is becoming increasingly common as the impacts of climate change and related processes mean that individuals and communities are required to struggle with the loss of "valued species, ecosystems and landscapes."[15] In this way, they carry death and extinction around with them as they go about their work. As Dave explained to me in the context of *Achatinella pupukanioe*: "I felt like I missed our window and I let potentially the last of a species slip through our fingers on my watch. . . . I've felt for the last couple of years like we're always five minutes too late."

I had already fallen in love with Hawai'i's snails when I discovered just how bad things were. I slowly came to realize, and accept, that there isn't really a "recovery plan" for these species in any full sense of this term. Of course, the absence of a plan is not for want of effort; it is simply a reflection of our current reality. And yet the more I have learned about the plight of Hawai'i's snails, the more convinced I have become of the importance of hope. The hope we need now, though—or at least the one we get, and need to learn to live well with—is a hope that is woven through with grief. Our hope must simultaneously be a work of care and mourning, a work that requires an honest reckoning with its own costs and dangers, and one that honors and acknowledges what has been and most likely will still be lost.

The lab, it seems to me, represents a vital site for this kind of hope. While mourning and grief are not often associated with hope, they carry within them the seeds of new understandings, relationships, and possibilities. The loss that can sometimes be disempowering, even debilitating, can also be an encounter that requires us to reevaluate our place in the world, to solidify our commitments, to change our practices. As the social scientists Neville Ellis and Ashlee Cunsolo have noted: "collective experiences of ecological grief may coalesce into a strengthened sense of love and commitment to the places, ecosystems and species that inspire, nurture and sustain us."[16] In this way we might move beyond a simplistic, binary understanding of the distinction between hope and grief to appreciate the way in which these responses can—and increasingly will—be intimately entangled with each other in our relationships with the living world around us.

The mournful hope we need in times like these is one grounded in the work of bearing witness. In the first instance, this witnessing is an act of faithfulness to individuals and species whose worlds we, collectively, have destroyed, to those snails already gone from the lab, like *Achatinella apex-fulva*, as well as those that didn't quite make it in, like *Achatinella pupukanioe*. But it is also an act of faithfulness to all of those that must now live out their lives in plastic containers, as well as to the larger species whose fates they carry with them.

In this context, even if species cannot ultimately be restored to the world beyond, holding onto them for as long as we can—provided that they are living flourishing lives—might be understood as an effort to cultivate some semblance of responsibility. After all, what is the alternative? To refuse to take responsibility at this time would be to inflict a double violence: to add our knowing indifference to the violence of carelessness and disregard that brought them to the edge of extinction.[17] This kind of indifference to extinction is one that, as the philosopher James Hatley has noted, wounds us, one that endangers our humanity.[18]

To bear witness is to refuse to turn away, to keep faith with these others. The anthropologist Deborah Bird Rose taught me about the obligations of witness in the face of extinction. She argued that to refuse to turn away was "to remain true to the lives within which ours are entangled, whether or not we can accomplish great change."[19] She saw this turning toward as "a willingness to situate one's self so as to be available to the call of others . . . a willingness toward responsibility, a choice for encounter and response."[20]

In its own powerful way, this witnessing is an act of hope. It is an act grounded in a hope that is at once more modest and more fundamental than any simple notion of success, a hope that, within the awful confines of the contemporary moment, we might cultivate the best modes of relationship and accountability with others that are still available to us.[21] This, it seems to me, is the very least that we owe to Hawai'i's snails and the countless other species living their lives at the edge of extinction.

But to bear witness is also to share this knowledge with others, to refuse to be silent in the face of death and extinction. At least in part, this act of sharing is grounded in the hope that it might make a difference, if not for those whose stories are told, then at least for others. This, too, is a mourning-work that the lab makes possible, and to some extent is already taking on. In contrast to the never-ending cycle of news stories in which extinctions, if they are noticed at all, gather a few headlines and are then quickly forgotten, the lab is a space of intensified and protracted loss. Of course, it is not only this. It is also a place of ongoing snail life and of intense care. But alongside these processes, tangled up with them, is the tragic reality of a growing group of tiny beings who stubbornly go on living even after their worlds have been so systematically destroyed.

The lab creates a site in which this reality can be encountered: by student groups, by hula and other cultural practitioners, by journalists, writers, and philosophers like myself. The lab creates the possibility for the kind of reflection that animates this book and travels out through it into the world. In this way, it helps to make visible, to share and testify to, an ongoing disappearance that might otherwise be much more difficult to see. In short, the lab creates the space for a public confrontation with loss: the losses of the past, those still to come, and those that we are often all too obliviously living through right now.

Without realizing it in these terms until I neared the end of writing, I now see that this book is itself an act of mournful hope in just this way. Emphatically, this does not mean that I have given up hope. Like everyone else I have met who has had the opportunity to get to know Hawai'i's remarkable snails, I am stubbornly holding onto the possibility that a full life in the forest might once again be possible for them. Even if we cannot see the path forward yet, we can work to hold open a clearing for it, which, in this case, means keeping the snails in the world in any way that we can. And yet the mourning cannot be lost sight of, either. This future will not arrive for all of these species, perhaps only for a few or even none of them.

The mournful hope that animates this book takes the form of an effort to cultivate a new and deeper sense of wonder and appreciation for snails and their worlds, of telling the stories that animate and inspire the search for flourishing possibilities for all. But it is also an effort to honor and acknowledge lives lived at the edge of extinction, facing up to and dwelling with the incredible processes of unraveling and wounding that are taking place here—and sharing those stories, too. And so, I tell these snail stories in the hope that they might be transformative of our present moment, of all of us. But also in the full knowledge that they may well not be—indeed, that it may already be too late—and yet with the conviction that telling them matters anyway.

EPILOGUE

We made our way along the rocky coastline under a baking sun. On the left of the narrow gravel path, the land dropped away down to the ocean. To our right, basalt cliffs rose up steeply above us. We were walking through a dry, scrubby landscape toward Ka'ena Point at the westernmost tip of O'ahu. I was with Brenden Holland again, this time in a very different environment. The heat in the day was palpable, radiating off every surface in this parched landscape. Very little rain falls on this leeward part of the island for most of the year, with the exception of a few winter months. Combined with the temperature, the salt spray, and the strong winds, this means that the only plants to be seen here are squat, mostly brown, grasses and low bushes. As we walked, the arid land around us sat in strange contrast to the cool waves that lapped against the shore. At one point, as I stopped to marvel at the rocks, Brenden turned back and laughingly noted: "We're a long way from Waikiki."

We had come to this rugged but beautiful coastline in search of traces of ancient snails. Of course, this is not the kind of environment in which one would normally expect to encounter moisture-dependent gastropods, but this place may not always have been this way. At least, that was the hypothesis that had brought us here.

About an hour into our walk, we rounded a corner and Brenden announced that we'd arrived at the spot. In front of us, on the mauka (inland) side of the path, a little land slip had brought sand and other debris down from the cliff above. It was here that we'd come to search for snails—or rather, for their shells. Right by the side of the path we began to sift through the sand with our hands, slowly making our way upward as we did. Almost immediately, we found what we were looking for, as Brenden knew we would, having visited this site many times before. There, scattered all over the place, were hundreds of little white shells. Most of them were conical in shape, and only about 6–7 millimeters in length. "They're from the family Amastridae," Brenden told me. "This one is the most common extinct species that we see out here."

Scattered among them were also the shells of two other disappeared snail species. One, a small flat shell about the same size, belonging to a species of the family Endodontidae; the other, even rarer, was a larger conical shell just over 1 centimeter in length, belonging to another species from the Amastridae. This latter shell, Brenden explained, is from a particularly interesting species, one of only two in the family with a sinistral, or left-spiraling, shell. As far as Brenden is aware, all three of these species are still among the unknown extinct, awaiting official description. Their shells, however, linger, and in large quantities—at least within this tiny spot. As Brenden noted, "you can walk for a mile and not see any and then suddenly there are hundreds."

In one of our earlier conversations, actually during the walk on Puʻu ʻŌhiʻa that I discussed in chapter 2, Brenden had mentioned these shells and I took the opportunity to impose on his hospitality, asking if he'd mind making a trip to show them to me. Having come across this spot several years earlier, Brenden had been captivated by this wild archive of ancient snail life. At the time of our visit, he was just beginning to plan some research on these shells that he hoped would shed a little light on the relationship between snails and a changing climate, helping us to understand the role of climate in both past and possible future gastropod extinctions in the islands.

For obvious reasons, much of our thinking about climate change–induced extinctions is focused on the present and future. But while the current period of anthropogenic climate change that we are already well within is going to usher in transformations on an unprecedented scale, the climate has also changed in the past. What lessons might snails offer us in understanding these processes and their impacts? As Brenden helpfully summed things up: "Snails make fantastic models for extinction . . . unfortunately. Mostly because they can't run away even though the environment is changing." Instead, they must either adapt or die out. When these changes are rapid, the latter is much more likely. Sadly, in these cases we discover that, as Brenden put it: "the only thing snails do fast is die."

Little research has been carried out to date on the deep, paleoclimatic history of the Hawaiian Islands. The work that has been done, however, indicates that on this leeward side of Oʻahu, the end of the last ice age about 11,000 years ago—which marks the start of the Holocene epoch—brought with it a period of warming and reduced rainfall, leading to significant changes in vegetation.[1] Even though this past climatic change happened relatively gradually, Brenden's hypothesis is that the ecosystem transformation that it brought about may have been too much for most of the resident snail species to bear.

This line of reasoning is grounded in a few key factors. Importantly, all of these species are members of larger families that still exist, and all of the other known members of these families live in very different environments to this one, specifically, in wet forests. "It's like night and day different," Brenden explained. This means that either these ancient snails were outliers in their respective families, inhabiting arid environments, and they went extinct in some other way, or the environment changed on them, perhaps bringing about their extinctions.

We certainly can't rule out the first of these possibilities, that these particular species were outliers. As Brenden noted in our conversation, there is another species of snail that can still be found in this area, *Succinea caduca*, and it is the only known member of its family in Hawaiʻi that is adapted to such an environment. But given what we know from other studies about shifts in Hawaiʻi's environment over the past several thousand years, Brenden thinks that climate change–induced extinction is more likely—or at least likely enough to warrant some further investigation.

To this end, he and his students have carbon-dated some of these ancient shells. While their ages vary, all of the shells tested so far belong to snails that lived sometime in the past 3,000–45,000 years. As such, they seem to have disappeared prior to human arrival, and during a period of ongoing climatic change. In the future, Brenden is planning to analyze the carbon 13 isotope signatures in the shells in the hope that this might tell us something about the plants that these snails lived among. In particular, these tests can provide a measure of the abundance of dry- versus wet-adapted plants in the environment.

Results from these tests can then be compared between snail shells from various time periods, perhaps revealing changes in vegetation. At the same time, the team will also conduct the same analyses on shells from close relative species, snails still living in the wet forests of Puʻu Hapapa. Their preliminary hypothesis is that the oldest of the shells from Kaʻena Point will have similar isotope signatures to these extant species—indicating a life among similar vegetation—while the youngest of the ancient snail shells, from around 3,000 years ago, will have inhabited a very different, much drier, environment and were perhaps among the last of their kind.

If these analyses all work out, this research might tell us something about these ancient snails while also enabling their shells to become a valuable archive for better understanding historical processes of environmental change, including nonanthropogenic extinction, in the islands. At the same time, of course, Brenden hopes that this situation might shed some further light on the potential tolerance and vulnerability of gastropods to these kinds of changes as we enter into a new period of growing climatic upheaval.

Whether or not it turns out that these ancient snail species succumbed to a changing climate, the fact remains that their contemporary relatives are facing a significant and growing challenge in this regard. Indeed, the biologist Melissa Price told me that the impacts of climate change may already be being felt by Hawaiʻi's snails. In the comparatively hot and dry environments found in parts of the Waiʻanae Range, she explained, we are seeing species disappearing from lower elevations. Past studies have shown that during drought conditions, juvenile snails suffer much higher rates of mortality than usual.[2] In her view, the current decline in snails seems to correlate with changes in the weather, as cloud cover is moving higher up within the valleys, reducing the available moisture in the forest.

Melissa explained that this view remains controversial, with others seeing predation as the almost exclusive factor driving these declines. To date there haven't been studies that have tried to map these overlapping patterns of snail and rainfall distribution, and given all the other impacts on snails that would need to be taken into account, this would be a highly complex modeling exercise. Based on what we know, however, Melissa told me that her hypothesis is that "climate change has reduced juvenile survival, which combined with heavy predation, has resulted in them disappearing from the lower elevations." In short, the snails are moving up the mountains with the clouds. Or, more accurately, they're dying out in those areas that are drying out.

While those snails that make their homes in the much wetter, windward Koʻolau Range on the other side of Oʻahu might be able to cope with a reduction in rainfall, in the already dry and highly variable Waiʻanae Range, even minor reductions might be catastrophic. This, however, is precisely what seems to be happening. Over the past several decades, a noticeable change has occurred in not only the amount of rainfall received in the Hawaiian Islands, but how that rain is being delivered. The cool northeast trade winds that prevail in the islands, and are relied on for their rains, have started to decline significantly. Recent studies have shown that, in Honolulu, for example, there were around 80 fewer days per year with these winds in 2009 when compared with weather patterns about four decades earlier.[3] This change correlates with a general decline in rainfall of roughly 15% in the two decades up to 2010.[4] Meanwhile, the amount of rain that falls during intense storm periods—which frequently cause landslides, flooding, and other hazards but do little good for ecosystems or to replenish the aquifers—has increased by roughly the same amount.[5]

It makes sense that snails would be among the first species to feel these changes. They are, after all, highly sensitive to moisture and temperature, and often have a relatively narrow band of tolerance for change. For this reason, snails—both in the oceans and on land—are increasingly being treated as "sentinel species." By monitoring their abundance and absence, we might, as the biologist Rebecca Rundell has put it, be able "to detect subtle changes that humans might not otherwise be able to see until it is too late."[6] This is something that, as Ken Hayes told me, is already starting to happen in Hawai'i, with various government agencies using standardized, regularly repeated snail counts to monitor the environments they're responsible for. In this way, snails are becoming a kind of "canary in the coal mine" for climate change—a worryingly apt metaphor for this period in which the dangers of extracting coal and other fossil fuels are becoming manifest in new ways.

Just how bad the impacts of climate change will be for Hawai'i's snails is something that no one is really able to predict with any certainty at this stage. It is, however, a looming presence, one that is increasingly demanding attention. When I spoke to Dave about snails and climate change in 2018, it was clear that he had other, much more pressing, matters on his mind. With so many populations being rapidly extinguished by predators, his focus was on evacuating them as quickly as possible. As he put it then, with an anxious laugh, climate change isn't really something that they have the luxury of thinking too much about. This is not to say that they were ignoring it. Indeed, wherever possible climate modeling has been being built into future planning for many years now, perhaps most significantly in the placement of new exclosures.

But things have shifted quickly on this front, too. While the pressure is very much still there to secure as many species as possible in the lab and the exclosures, and the impacts of predators have, if anything, increased, so have the impacts of climate change. When I spoke to Dave again about this topic in 2020, he told me that the SEPP team was contemplating the very real prospect that in the next 60–100 years, the climate within the historical ranges of many of O'ahu's snail species may no longer be suitable for them at all. As such, in order to survive, they might need to be relocated elsewhere.

For some snail species, on some islands, it may be possible to "simply" move them up the mountainside. But once they reach the upper elevations of their respective mountains, this option is obviously no longer possible. In these cases, it may be necessary to consider translocating species outside their historical ranges, to another mountain or perhaps an entirely different island. At the very least, Dave, Melissa, and others think that these are possibilities that we need to start considering now. This prospect raises a range of challenges, from legislative obstacles surrounding the movement of endangered species and risks of hybridization with other species to community objections. None of these barriers is insignificant, but in the end, if the climate keeps changing, it may be the only option available for life outside an environmental chamber.

At least for the foreseeable future, it is highly likely that any snails that were translocated to new environments would need to have an exclosures built for them in their new home. There are no Hawaiian islands free of wolfsnails that might provide a haven, and there is no talk of moving snails beyond the archipelago. It is pretty unlikely that any species translocated in this way would establish itself outside the exclosure it was moved to. As Dave noted, while "snails can get out of the exclosure—they tend to get eaten when they do." In this way, the existing and growing network of exclosures might become a home to even more species, and novel combinations of species. If a solution is one day found to the threat of predators, these new landscapes might also be the most appropriate environments to release these snails into. As we have seen, however, such a future is a long way off—if it arrives at all.

Hawai'i's snails are by no means alone here. All over the islands, and increasingly around the world, conservationists are beginning to wonder about—and in some cases to implement—programs of translocation to prepare plants and animals for our shifting climatic reality. In fact, just a little further along the path that Brenden and I were walking on at Ka'ena Point, there is a large, fenced area that has been created with precisely this future in mind. Throughout the north Pacific Ocean, many seabirds, including the majority of some endangered species like the Laysan and black-footed albatrosses, nest on low-lying islands that are already experiencing the impacts of rising sea levels, increased storm surges, and saltwater inundation. In an effort to safeguard these species, conservationists have worked to establish new breeding colonies on higher islands like O'ahu. While at Ka'ena Point these species were not moved outside their historical range, just bolstered within it, in other cases it has been necessary to look to entirely new environments. Some conservationists are arguing that, as the years go by, this is likely to become an increasingly necessary practice.[7]

Admittedly, snails are not at the top of most lists for expensive conservation translocations. If, however, it does become necessary and possible to relocate some of Hawai'i's snails to more appropriate environments, then, somewhat perversely, we might end up looking back on this period of snail evacuations and being thankful for the lessons it has provided. All the work that has been and is being done in the lab, the exclosures, the museum, and elsewhere has provided invaluable information about how to keep snails alive in captivity, how to relocate them, and how to deal with challenges like their wandering. If snail translocations do become necessary, perhaps on a significant scale, then this background might be one that ends up saving species.

None of this is to say that it isn't worth fighting to maintain as much of our current environments and climatic conditions as we possibly can. Moving species around the world, even if it can be done well, will never provide ideal conservation outcomes. Rather, it is and must remain an approach of last resort.[8] As such, the work of holding onto snails in exclosures and the lab has to be coupled with broader action on climate change and forest conservation. It is only in this way that, should a solution to the challenge of wolfsnail predation one day emerge, Hawai'i's snails might still be able to return to their ancestral forests, to those particular landscapes that they have evolved within and helped to shape through their millions of years of journeying through the leaves.

The specter of climate change is a grave presence to invoke in the closing pages of this book. And yet there is no way of getting around the fact that it is one that we will increasingly be required to live with. In this, and so many other ways, I don't know what the future holds for Hawai'i's snails. No one does. But in the meantime, many of the passionate people and groups whose stories I have shared in this book are getting on with the work of holding open and actively crafting room in the world for them.

At the Bishop Museum, the triage taxonomy continues. In 2020, Nori, Ken, and their colleagues published another new Hawaiian snail species description: *Auriculella gagneorum*. This one, however, managed to grab a few media headlines. While numerous species of already extinct snails have been discovered in recent decades, this description is for a species that is still hanging on.[9] On the one hand, this situation might be read as a reminder of the need for more investment in invertebrate taxonomy: the first specimens of this species were collected roughly 100 years ago, and even then, they were labeled by Yoshio Kondo "NSP," indicating a new species. It has taken this long for Kondo's assessment to be officially confirmed. On the other hand, however, as Nori reminded me when I asked her about this work, this new species is a reminder of how much is still out there. Even today, roughly 10% of the Hawaiian land snail specimens they collect in the field are thought to be from undescribed species. While much has been lost, much more than we will ever know, it turns out that there is also much more than we have yet appreciated that can still be held onto.

Meanwhile, Dave and the SEPP team are expanding their operations, with one permanent staff member and a growing network of exclosures on Maui, and a new collaboration with private landowners that has led to the construction of two exclosures on Lānaʻi. On Oʻahu, too, the network of exclosures is expanding quickly. If everything goes to plan, a couple of years from now almost as many exclosures will have been built in a 5-year period as were built in the previous two decades. At the same time, funding has been obtained to double the size of the lab. This new facility should start welcoming lodgers in 2022 and will one day be home to over 10,000 snails. Not content with this, though, Dave has been hard at work securing funds to create duplicate lab facilities with partners at the Bishop Museum and the Honolulu Zoo. As he explained to me: "This is a really neat partnership that will allow us to spread vulnerable populations between the three facilities to reduce the risk of something stochastic causing an extinction in one facility."

These developments are part of a new energy around the conservation of snails in Hawaiʻi. Within the scientific community, a few decades ago only a handful of people really knew and cared about them. Today, it is possible to assemble a group of more than 40 biologists whose sole or primary focus is the snails of these islands. In fact, in early 2020, this assembly took place: Hui Kāhuli, a full-day event at the museum. When I chatted to Dave about the meeting a few months later, he told me that this would never have been possible even 10 years ago. But something is changing; a new momentum is gathering. "We're building on the foundation that Mike established with the captive rearing program all those years ago. My hope is that this momentum will be enough to carry us all through."

Importantly, all of this snail-focused work takes place within the context of a growing Kanaka Maoli movement against the destruction of the 'āina and the communities of human and more-than-human life that it supports. As is happening in so many parts of the world today, the Indigenous people of these islands are fighting back against the ravages of climate change, capitalism, militarism, and extraction, and working to make visible the deep-seated connections between these processes and colonization.[10] Often, as Noelani Goodyear-Ka'ōpua has noted, this resistance is one that seeks to simultaneously underscore "the ways that imperialist industrial projects harm Indigenous Pacific cultures," while also drawing on "those very cultural practices of renewing connections with lands and waters in order to engage in direct action struggle."[11] Through this struggle, Kānaka Maoli and their allies are working to unsettle some of the key processes that have driven the past declines of the islands' snails and so many other animal and plant species, processes that are today ushering in the transformed climate that is perhaps the snails' greatest emerging threat.

Alongside these efforts, a growing group of scientists and cultural practitioners—many of whom move across and between these knowledge traditions—are working to root conservation efforts in Kanaka Maoli wisdom and practices. As Kawika Winter, Kamanamaikalani Beamer, Mehana Blaich Vaughan, and their colleagues have argued, in the islands today there is "a growing recognition of the ingenuity of Hawaiian biocultural resource management systems. These systems effectively adapted to local conditions, while accumulating a body of knowledge in response to observed effects of management—both successes and failures—in order to sustain resource abundance over time."[12] Across diverse contexts, from conservation action on the ground to educational initiatives, the mutually sustaining and nourishing connection between Kanaka Maoli culture and the 'āina is being acknowledged and enacted.[13] In this way, practitioners hope that the processes of biocultural loss and damage that have rippled across these domains throughout the long history of colonization in these islands might be turned around to create new opportunities for mutual resurgence. Here, too, the snails have a role to play. As Sam 'Ohu Gon succinctly summed up the situation in one of our snail conversations: "Any time you find a plant or animal that has cultural significance it is an important thing, because it is not only a motivation for the preservation of that species, it is a motivation for the re-spark and reconnection of Hawaiian culture back to the natural world."

This book has aimed to show and explore how stories might be a vital aspect of this effort to hold onto snails and other disappearing species while acknowledging, resisting, and even redoing the larger processes that are driving these losses. As we enter ever more deeply into our planet's sixth mass extinction event, into a period of unfathomable losses both seen and unseen, losses that ripple out across diverse lives and landscapes, storytelling takes on ever more importance in enabling us to creatively, inclusively, and effectively imagine and craft alternatives to our current trajectory.

Above all else, this book has aimed to add some flesh to our sense of what is lost in extinction. It is too easy to dismiss the disappearance of yet another snail species, another group of insignificant, slimy, little creatures. Attending to their stories, learning to see and hear snails differently, draws us into other kinds of appreciation for the significance of this loss. Hawai'i's snails carry with them diverse modes of life: ways of navigating and interpreting the world, ways of socializing and of feeding, rich even if far from intelligible modes of being. They also carry immense evolutionary lineages, deep histories of ocean crossings, of journeys by bird or log, and then of movements within these island landscapes that have enabled one of the planet's most diverse radiations of snail species to evolve. In this way these snails moved into, and helped to shape, the ecological communities of these islands as they cleaned and decomposed leaves. They also helped to produce soils and healthy forests—or, at the very least, there is a good chance that they once did, to what extent, in what ways, and with what consequences, we will probably never really know.

These snails also carry vibrant and vital cultural relationships. They do this in particular for Kānaka Maoli, whose moʻolelo and mele they have woven their way into, becoming part of a vibrant more-than-human world of kinship and care. While many of these relationships have been damaged by more than a century of colonization, militarization, deforestation, and more, the work of (re)connection and of resistance to destruction—the work of aloha ʻāina—is ongoing in Mākua Valley and across these islands. And, as we have seen, snails have been and might still be powerful allies in this work, too.

We need stories to draw out this complexity, these connections, histories, and possibilities. In particular, we need stories that attend to the simultaneously biological and cultural reality of extinctions, patterns of unraveling and remaking that cut across any neat divide between what are customarily the domains of the natural sciences and the humanities. In this way we are able to grapple with the particularity of extinctions, with their diverse, uneven, and highly consequential forms.

And so, while we have seen that snails do not really carry their homes on their backs—rather, they craft them with other snails, writing them into the world through their own slime trails and accumulated experiences—they do carry a great deal else. All of this is at stake in extinction. Each and every extinction has its own story. Or, rather, each is its own complex set of nested and interwoven stories, reaching into the deep past but ongoing, drawing in diverse living beings in different ways. In short, contra the title of this book, we see that there are *multiple* worlds both tangled up, and at stake, in a shell.

Back along the rocky coastline of Kaʻena Point, as I sifted through the sand for shells, turning over in my hands the lingering traces of so many long-lost lives and ways of life, my mind turned to the forest. I thought back to those first kāhuli that I had encountered among the trees, quietly estivating on a piece of fluorescent pink tape, safely tucked away behind an exclosure fence. I wondered if they were there again now, at the other end of this same mountain range, among the clouds. I wondered if they would soon set off on their nightly adventures up into the branches, laying down slimy trails for themselves and others, as has been the way of their species in these islands for millions of years. And I hoped. I hoped that 100 years from now, even 100,000, the ancient shells that I now held in my hand wouldn't be the only kind of trace that is left of their lives. And I dared, for just a moment, to hope an even grander hope: that one day, when all is pono—when all is as it should be in the forest once more—the descendants of those same snails might be able to leave the confines of their current home, to spread out across this mountain range, and to fill the night air with their wondrous song.

Acknowledgments

This book would not have been possible without the support and generosity of the many people in Hawai'i who guided my journey into the remarkable worlds of snails. It was these people, as much as the snails themselves, who drew me to want to tell these stories. They patiently answered my many questions as well as reading and commenting on chapter drafts. More than anything else, though, they shared with me their love of snails. My sincere thanks to Mike Hadfield, Dave Sischo, Nori Yeung, Ken Hayes, Brenden Holland, Rob Cowie, and Carl Christensen.

Many other biologists and conservationists in Hawai'i and beyond also agreed to be interviewed and talk about snails with me over the years that I was researching this book. Thanks to Melissa Price, Lindsay Renshaw, Vince Costello, Sarah Dalesman, Ken Lukowiak, Ronald Chase, and Jon Ablett.

A diverse group of people also generously gave of their time and their insight to share with me the many ways in which snails have woven themselves into the lives and culture of Kānaka Maoli. These people are kumu hula and kumu oli, they are teachers and protectors of 'ōlelo Hawai'i, they are scientists, activists, artists, and scholars. My sincere thanks to Uncle Sparky Rodrigues, Aunty Lynette Cruz, Cody Pueo Pata, Puakea Nogelmeier, Sam 'Ohu Gon III, Larry Lindsey Kimura, Kupa'a Hee, Kyle Kajihiro, Vince Dodge, and Justin Hill. Thank you also to David Henkin for sharing his many insights into the legal struggles over Mākua Valley.

This book was also greatly enriched by several important academic contexts, including colleagues who discussed ideas and commented on chapter drafts or presentations. My thanks to the staff and students of the Center for Pacific Islands Studies at the University of Hawaiʻi at Mānoa who hosted me during my longest research visit. Particular thanks to Alex Mawyer, who provided rich and generous feedback, as well as ongoing support and encouragement. Many other people read drafts or discussed key ideas with me. My particular thanks to Warwick Anderson, Michelle Bastian, Brett Buchanan, Danielle Celermajer, Sophie Chao, Matthew Chrulew, Vinciane Despret, Donna Haraway, Julia Kindt, Eben Kirksey, Britt Kramvig, Hélène Ahlberger Le Deunff, Jamie Lorimer, Stephen Muecke, Ursula Münster, Brandy Nālani McDougall, Myles Oakey, Emily O'Gorman, Craig Santos Perez, Elspeth Probyn, Hugo Reinert, David Schlosberg, Isabelle Stengers, Heather Swanson, Anna Tsing, Jane Ulman, Jamie Wang, and Sam Widin. A particular thanks to Donna Haraway, who, unknowingly, I think, initiated this project when she introduced me to her friend Mike Hadfield, and so to the snails, almost 10 years ago now.

Thank you also to the various scholars who provided (mostly anonymous) feedback on the manuscript as formal reviewers for the press. A particular thanks to Candace Fujikane, whose insightful commentary added new layers to the discussion in several key places.

The map for this book was created by Richard Morden. My thanks to him for his careful work.

Finally, my sincere thanks to Beth Clevenger at the MIT Press for her unwavering support and enthusiasm for this project. It has been an absolute pleasure working with you and the rest of the team, including Anthony Zannino, Deborah Cantor-Adams, Stephanie Sakson, Molly Seamans, Mary Reilly, and Jay Martsi.

Funding for this research was provided by the Australian Research Council (FT160100098) and the University of Sydney.

Sections of some of the chapters in this book have been published in abridged or edited form in the following places: T. van Dooren, "Snail Trails: A Foray into Disappearing Worlds, Written in Slime," in Sarah Bezan and Robert McKay (eds.), *Animal Remains* (London: Routledge, 2021); T. van Dooren, "The Disappearing Snails of Hawaiʻi: Storytelling for a Time of Extinctions," in Thom van Dooren and Matthew Chrulew (eds.), *Kin: Thinking with Deborah Bird Rose* (Durham, NC: Duke University Press, 2022); T. van Dooren, "Drifting with Snails: Extinction Stories from Hawai'i," in Kaori Nagai (ed.), *Maritime Animals* (University Park: Penn State University Press, 2022); T. van Dooren, 2022, "In Search of Lost Snails: Storying Unknown Extinctions," *Environmental Humanities*, 14.1; T. van Dooren, "Military Snails: Multispecies Solidarities in Hawai'i," under review; T. van Dooren, "Hospice Earth," *Overland*, 2019; T. van Dooren, "Mourning as Care in the Snail Ark," in Ursula Münster, Thom van Dooren, Sara Asu Schroer, and Hugo Reinert (eds), *Multispecies Care in the Sixth Extinction*, an online collection (Society for Cultural Anthropology, 2021).

Glossary

HAWAIIAN TERMS

ahupuaʻa	a traditional land division that often runs from the mountains to the sea
ʻāina	land; literally, "that which nourishes"
akua	deities
aliʻi	chief
hālau hula	a school of hula
hōʻailona	omen, sign
kaona	a hidden or inner meaning/reference that is not apparent to all
kāhuli	snail
kalo	taro (plant)
kanaka	native; literally, "human being" or "person"
Kanaka Maoli	native Hawaiian (plural: Kānaka Maoli)
keiki	child
kīpuka	a section of forest that is left standing after a lava flow moves through a landscape. It is from these preserved areas that seeds, spores, and other forms of life are able to spread out, to reforest the landscape. The term is used more widely to refer to places of refuge and sources of regeneration.

kuahu	altar
kuleana	a right/privilege and a responsibility
kumu	teacher, source
kupuna	elder (plural: kūpuna)
makaʻāinana	common people; literally, "people who attend the land"
mālama	care
moʻolelo	story (including history)
mele	traditional song or chant
ʻōlelo Hawaiʻi	Hawaiian language
pono	balanced, harmonious, proper
pūpū	shell; used to refer to shelled creatures and to their shells, both terrestrial and aquatic
ʻuala	sweet potato

SCIENTIFIC TERMS AND ACRONYMS

aperture	the main opening of a snail's shell from which the head-foot of the animal emerges
biogeography	the study of the distribution of species across geographical space and across evolutionary time
conspecific	belonging to the same species
EIS	Environmental Impact Statement
ESA	US Endangered Species Act
estivation	rest/dormancy during a hot and/or dry period
gastropod	a member of the taxonomic group comprising the snails and slugs
invertebrate	an animal without a backbone
IUCN	International Union for Conservation of Nature
malacology	the scientific study of mollusks

mollusk	a large taxonomic group (phylum) that includes everything from clams, mussels, snails, and slugs to octopuses and squid
NEPA	National Environmental Policy Act
OANRP	Oʻahu Army Natural Resources Program
radula	a specialized appendage used by snails for feeding, essentially a tongue that is lined with hundreds of tiny teeth
SEPP	Hawaiʻi Snail Extinction Prevention Program
taxonomy	the scientific description and classification of species
Umwelt	the world of meaning/experience inhabited and crafted by an organism (plural: *Umwelten*)
USFWS	US Fish and Wildlife Service

Notes

INTRODUCTION

1   For a fuller discussion of the ongoing research on the dating of Polynesian arrival in the Hawaiian Islands, see Patrick V. Kirch, "When Did the Polynesians Settle Hawai'i? A Review of 150 Years of Scholarly Inquiry and a Tentative Answer," *Hawaiian Archaeology* 12 (2011).

2   Sam 'Ohu Gon and Kawika Winter, "A Hawaiian Renaissance That Could Save the World," *American Scientist* 107, no. 4 (2019).

3   David D. Baldwin, "The Land Shells of the Hawaiian Islands," *Hawaiian Almanac and Annual* (1887): 62.

4   Norine W. Yeung and Kenneth A. Hayes, "Biodiversity and Extinction of Hawaiian Land Snails: How Many Are Left Now and What Must We Do to Conserve Them—A Reply to Solem (1990)," *Integrative and Comparative Biology* 58, no. 6 (2018).

5   Alan D. Hart, "Living Jewels Imperiled," *Defenders* 50 (1975).

6   Richard Primack, *Essentials of Conservation Biology* (Sunderland, MA: Sinaur Associates Inc., 1993); Anthony D. Barnosky et al., "Has the Earth's Sixth Mass Extinction Already Arrived?," *Nature* 471 (2011).

7   IUCN Summary Statistics. "Table 3: Species by Kingdom and Class." Available online: https://www.iucnredlist.org/statistics. These and all subsequent figures refer to Version 2021-1 of the Red List.

8   Ronald B. Chase, *Behavior and Its Neural Control in Gastropod Molluscs* (Oxford: Oxford University Press, 2002), 3.

9   David Sepkoski, *Catastrophic Thinking: Extinction and the Value of Diversity from Darwin to the Anthropocene* (Chicago: University of Chicago Press, 2020), 24–27.

10   C. Mora et al., "How Many Species Are There on Earth and in the Ocean?," *PLoS Biology* 9, no. 8 (2011).

11 Robert H. Cowie et al., "Measuring the Sixth Extinction: What Do Mollusks Tell Us?," *Nautilus* 1 (2017).

12 Yeung and Hayes, "Biodiversity and Extinction of Hawaiian Land Snails."

13 Alan Solem, "How Many Hawaiian Land Snail Species Are Left? And What We Can Do for Them?," *Bishop Museum Occasional Papers* 30 (1990); Yeung and Hayes, "Biodiversity and Extinction of Hawaiian Land Snails."

14 Figures shared at Hui Kāhuli meeting organized by Dave Sischo, Bishop Museum, Honolulu, HI, March 9, 2020.

15 Among the land snails, they are the nine remaining species (of 41) of the genus *Achatinella* on Oʻahu, alongside one species from the island of Maui and two from Lānaʻi.

16 https://www.hawaiibusiness.com/endangered-and-underfunded/. Also see David L. Leonard Jr., "Recovery Expenditures for Birds Listed under the US Endangered Species Act: The Disparity between Mainland and Hawaiian Taxa," *Biological Conservation* 141 (2008).

17 Yeung, and Hayes, "Biodiversity and Extinction of Hawaiian Land Snails," 1163.

18 Timothy R. New, "Angels on a Pin: Dimensions of the Crisis in Invertebrate Conservation," *American Zoologist* 33, no. 6 (1993). Also see Pedro Cardoso et al., "The Seven Impediments in Invertebrate Conservation and How to Overcome Them," *Biological Conservation* 144, no. 11 (2011).

19 Edward O. Wilson, "The Little Things That Run the World (the Importance and Conservation of Invertebrates)," *Conservation Biology* 1, no. 4 (1987). Also see C. M. Prather et al., "Invertebrates, Ecosystem Services and Climate Change," *Biological Reviews of the Cambridge Philosophical Society* 88, no. 2 (2013).

20 Aimee You Sato, Melissa Renae Price, and Mehana Blaich Vaughan, "Kāhuli: Uncovering Indigenous Ecological Knowledge to Conserve Endangered Hawaiian Land Snails," *Society & Natural Resources* 31, no. 3 (2018): 324.

21 Thom van Dooren, *Flight Ways: Life and Loss at the Edge of Extinction* (New York: Columbia University Press, 2014); Thom van Dooren and Deborah Bird Rose, "Lively Ethography: Storying Animist Worlds," *Environmental Humanities* 8 (2016).

22 Brett Buchanan, Michelle Bastian, and Matthew Chrulew, "Introduction: Field Philosophy and Other Experiments," *Parallax* 24, no. 4 (2018).

23 For some great examples of environmental popular science and nature writing that grapple with processes of colonization, globalization, militarization, and dispossession, see Robin Wall Kimmerer, *Braiding Sweetgrass: Indigenous Wisdom, Scientific Knowledge and the Teachings of Plants* (Minneapolis, MN: Milkweed Editions, 2013); Raja Shehadeh, *Palestinian Walks: Notes on a Vanishing Landscape* (London: Profile Books, 2008); Michelle Nijhuis, *Beloved Beasts: Fighting for Life in an Age of Extinction* (New York: W. W. Norton, 2021); Rebecca Giggs, *Fathoms: The World in the Whale* (London: Scribe, 2020); Amitav Ghosh. *The Nutmeg's Curse: Parables for a Planet in Crisis* (Chicago: University of Chicago Press,

2021), as well as the extensive bodies of work on food and agriculture by both Vandana Shiva and Gary Paul Nabhan. There is also a growing body of scholarly work on the cultural and philosophical dimensions of species extinction; see Deborah Bird Rose and Thom van Dooren, eds., "Unloved Others: Death of the Disregarded in the Time of Extinctions," *Australian Humanities Review* 50 (2011); Deborah Bird Rose, Thom van Dooren, and Matthew Chrulew, *Extinction Studies: Stories of Time, Death and Generations* (New York: Columbia University Press, 2017); Deborah Bird Rose, *Wild Dog Dreaming: Love and Extinction* (Charlottesville: University of Virginia Press, 2011); van Dooren, *Flight Ways*; Ursula K. Heise, *Imagining Extinction: The Cultural Meanings of Endangered Species* (Chicago: University of Chicago Press, 2016); Susan McHugh, *Love in a Time of Slaughters: Human–Animal Stories against Genocides and Extinctions* (University Park: Penn State University Press, 2019); Dolly Jørgensen, *Recovering Lost Species in the Modern Age: Histories of Longing and Belonging* (Cambridge, MA: MIT Press, 2019); and Jamie Lorimer, *Wildlife in the Anthropocene: Conservation after Nature* (Minneapolis: University of Minnesota Press, 2015).

24 Noel Castree, "The Anthropocene and the Environmental Humanities: Extending the Conversation," *Environmental Humanities* 5 (2014); Nigel Clark, "Geo-Politics and the Disaster of the Anthropocene," *The Sociological Review* 62 (2014); Heather Davis and Zoe Todd, "On the Importance of a Date, or Decolonizing the Anthropocene," *ACME: An International E-Journal for Critical Geographies* 16, no. 4 (2017); Elizabeth M. DeLoughrey, *Allegories of the Anthropocene* (Durham, NC: Duke University Press, 2019); Lesley Head, "The Anthropoceneans," *Geographical Research* 53, no. 3 (2015); Elizabeth Johnson et al., "After the Anthropocene: Politics and Geographic Inquiry for a New Epoch," *Progress in Human Geography* 38, no. 3 (2014); and Kyle Powys Whyte, "Our Ancestors' Dystopia Now: Indigenous Conservation and the Anthropocene," in *The Routledge Companion to the Environmental Humanities*, ed. Ursula K. Heise, Jon Christensen, and Michelle Niemann (London: Routledge, 2016).

25 Ashley Dawson, *Extinction: A Radical History* (New York: OR Books, 2016), 88.

26 Michael Hadfield, "Snails That Kill Snails," in *The Feral Atlas*, ed. Anna L. Tsing, Jennifer Deger, Alder Keleman Saxena, and Feifei Zhou (Stanford, CA: Stanford University Press, 2021).

27 Jonathan Kay Kamakawiwoʻole Osorio, *Dismembering Lāhui: A History of the Hawaiian Nation to 1887* (Honolulu: University of Hawaiʻi Press, 2002), 3.

28 This situation reminds us that, while the Anthropocene might mark a moment of radical upheaval—perhaps even the end of the world—for many people, especially Indigenous communities, "imperialism and ongoing (settler) colonialisms have been ending worlds for as long as they have been in existence." Kathryn Yusoff, *A Billion Black Anthropocenes or None* (Minneapolis: University of Minnesota Press, 2018). Also see Davis and Todd, "On the Importance of a Date."

29 Ty P. Kāwika Tengan, *Native Men Remade: Gender and Nation in Contemporary Hawaiʻi* (Durham, NC: Duke University Press, 2008), 67.

30 Noelani Goodyear-Kaʻōpua, "Protectors of the Future, Not Protestors of the Past: Indigenous Pacific Activism and Mauna a Wākea," *South Atlantic Quarterly* 116, no. 1 (2017).

31 To paraphrase and slightly redo one of the core questions posed by the feminist science studies scholar Donna Haraway. Donna Haraway, *When Species Meet* (Minneapolis: University of Minnesota Press, 2008).

32 Audra Mitchell, "Revitalizing Laws, (Re)-Making Treaties, Dismantling Violence: Indigenous Resurgence against 'the Sixth Mass Extinction,'" *Social & Cultural Geography* 21, no. 7 (2020).

33 Henry A. Pisbry and C. Montague Cooke, *Manual of Conchology*, vol. 22 (Philadelphia: Academy of Natural Sciences, 1912), 320.

34 As Sarah Bezan has noted, endlings can compress and concretize "macrohistorical processes of extinction that, due to their scale and complexity, evade full comprehension." In so doing, they personalize an abstract process in the lives of particular individuals. Indeed, as Bezan argues, when endling stories are told they frequently recount the lives and deaths of those last animal individuals and of the humans who cared for them. Sarah Bezan, "The Endling Taxidermy of Lonesome George: Iconographies of Extinction at the End of the Line," *Configurations* 27, no. 2 (2019). On this topic, also see Dolly Jørgensen, "Endling, the Power of the Last in an Extinction-Prone World," *Environmental Philosophy* 14, no. 1 (2017).

CHAPTER 1

1 Peter Williams, *Snail* (London: Reaktion Books, 2009), 14.

2 Brett Buchanan, "On the Trail of a Philosopher Ethologist," paper presented at The History, Philosophy, and Future of Ethology IV, Curtin University, Perth, 2019.

3 Ben Woodard, *Slime Dynamics* (Winchester, UK: John Hunt Publishing, 2012).

4 Martha Warren Beckwith, trans., *The Kumulipo: A Hawaiian Creation Chant* (Chicago: University of Chicago Press, 1951). On the *Kumulipo*, see Brandy Nālani McDougall, *Finding Meaning: Kaona and Contemporary Hawaiian Literature* (Tucson: University of Arizona Press, 2016).

5 Alan D. Hart, "The Onslaught against Hawaii's Tree Snails," *Natural History* 87 (1978); USFWS, "Endangered and Threatened Wildlife and Plants; Listing the Hawaiian (Oahu) Tree Snails of the Genus *Achatinella*, as Endangered Species," *Federal Register* 46 FR 3178 (1981).

6 For a fuller discussion of the diversity and distribution of snail species across the Hawaiian archipelago, see Robert H. Cowie, "Variation in Species Diversity and Shell Shape in Hawaiian Land Snails: *In Situ* Speciation and Ecological Relationships," *Evolution* 49, no. 6 (1995).

7 Michael G. Hadfield and Barbara Shank Mountain, "A Field Study of a Vanishing Species, *Achatinella mustelina* (Gastropoda, Pulmonata), in the Waianae Mountains of Oahu," *Pacific Science* 34, no. 4 (1980); Hadfield, "Snails That Kill Snails."

8 Michael G. Hadfield and Donna Haraway, "The Tree-Snail Manifesto," *Current Anthropology* 60, no. S20 (2019).

9 Wallace M. Meyer et al., "Two for One: Inadvertent Introduction of *Euglandina* Species during Failed Bio-Control Efforts in Hawaii," *Biological Invasions* 19, no. 5 (2017).

10 C. J. Davis and G. D. Butler, "Introduced Enemies of the Giant African Snail, *Achatina fulica* Bowdich, in Hawaii (Pulmonata: Achatinidae)," *Proceedings of the Hawaiian Entomological Society* 18, no. 3 (1964). Also see Robert H. Cowie, "Patterns of Introduction of Non-Indigenous Non-Marine Snails and Slugs in the Hawaiian Islands," *Biodiversity and Conservation* 7, no. 3 (1998).

11 Hadfield, "Snails That Kill Snails."

12 Meyer et al., "Two for One."

13 Thom van Dooren, "Invasive Species in Penguin Worlds: An Ethical Taxonomy of Killing for Conservation," *Conservation and Society* 9 (2011).

14 Claire Régnier, Benoît Fontaine, and Philippe Bouchet, "Not Knowing, Not Recording, Not Listing: Numerous Unnoticed Mollusk Extinctions," *Conservation Biology* 23, no. 5 (2009): 1218.

15 In my conversations with snail biologists, I've discovered that the term "estivate" is used in different ways. While some reserve it for an extended period of summertime/hot season inactivity (in effect, the opposite of a winter "hibernation"), others use it to describe a broad range of inactive, restful, periods—including those that many snails engage in during the warm daylight hours. It is in this broader sense that I have used the term in this book.

16 Chase, *Behavior and Its Neural Control in Gastropod Molluscs*, 34–52.

17 A. Cook, "Homing by the Slug *Limax pseudoflavus*," *Animal Behaviour* 27 (1979).

18 T. P. Ng et al., "Snails and Their Trails: The Multiple Functions of Trail-Following in Gastropods," *Biological Reviews* 88, no. 3 (2013): 684.

19 Mark Denny, "The Role of Gastropod Pedal Mucus in Locomotion," *Nature* 285 (1980).

20 Ng et al., "Snails and Their Trails," 685.

21 Denny, "The Role of Gastropod Pedal Mucus in Locomotion"; Ng et al., "Snails and Their Trails," 684.

22 M. S. Davies and J. Blackwell, "Energy Saving through Trail Following in a Marine Snail," *Proceedings of the Royal Society B* 274 (2007).

23 Jakob von Uexküll, *A Foray into the Worlds of Animals and Humans: With a Theory of Meaning*, trans. Joseph D. O'Neil (Minneapolis: University of Minnesota Press, 2010).

24 In turning to this work on slime trails it is not lost on me that, somewhat ironically and no doubt misleadingly, it may well be the fact that snails leave a *visible* trace of their chemicals in the form of mucus that makes their chemosensory worlds *seem* a little more accessible and intelligible to vision-centric creatures like us than those of so many other organisms.

25 Vinciane Despret, "The Enigma of the Raven," *Angelaki* 20 (2015).

26 Joris M. Koene and Andries Ter Maat, "Coolidge Effect in Pond Snails: Male Motivation in a Simultaneous Hermaphrodite," *BMC Evolutionary Biology* 7, no. 1 (2007).

27   Ng et al., "Snails and Their Trails," 689.

28   Carl Edelstam and Carina Palmer, "Homing Behaviour in Gastropodes," *Oikos* 2 (1950).

29   Edelstam and Palmer, "Homing Behaviour in Gastropodes," 266.

30   Ng et al., "Snails and Their Trails," 691.

31   This is not a competing story. The fact that snails might enjoy and seek out (some) others' company provides a psychological or physiological motivation for this behavior. Functional and motivational explanations operate on different biological levels. Both can be true at the same time, depending on the questions we're interested in. In fact, they may well reinforce one another: for example, the fact that snails *like* to aggregate may be one of the evolved mechanisms through which this functional advantage is achieved.

32   This research has focused on the pond snail (*Lymnaea stagnalis*), not because they are thought to have particularly complicated or fascinating capacities in this area, but rather because their neural anatomy is relatively simple and therefore easier to study.

33   K. Lukowiak et al., "Environmentally Relevant Stressors Alter Memory Formation in the Pond Snail *Lymnaea*," *Journal of Experimental Biology* 217, no. 1 (2014).

34   Brenden S. Holland et al., "Tracking Behavior in the Snail *Euglandina rosea*: First Evidence of Preference for Endemic vs. Biocontrol Target Pest Species in Hawaii," *American Malacological Bulletin* 30, no. 1 (2012).

35   Kavan T. Clifford et al., "Slime-Trail Tracking in the Predatory Snail, *Euglandina rosea*," *Behavioral Neuroscience* 117, no. 5 (2003).

36   Wallace M. Meyer and Robert H. Cowie, "Feeding Preferences of Two Predatory Snails Introduced to Hawaii and Their Conservation Implications," *Malacologia* 53, no. 1 (2010).

37   Nagma Shaheen et al., "A Predatory Snail Distinguishes between Conspecific and Heterospecific Snails and Trails Based on Chemical Cues in Slime," *Animal Behaviour* 70, no. 5 (2005); Clifford et al., "Slime-Trail Tracking in the Predatory Snail, *Euglandina rosea*."

38   Gary Snyder, *The Practice of the Wild* (Berkeley, CA: Counterpoint Press, 2010), 120.

39   Carlo Brentari, *Jakob von Uexküll: The Discovery of the Umwelt between Biosemiotics and Theoretical Biology* (Dordrecht: Springer, 2015), 240–241. Also see Matthew Chrulew, "Reconstructing the Worlds of Wildlife: Uexküll, Hediger, and Beyond," *Biosemiotics* 13, no. 1 (2020).

40   Brenden S. Holland, Marianne Gousy-Leblanc, and Joanne Y. Yew, "Strangers in the Dark: Behavioral and Biochemical Evidence for Trail Pheromones in Hawaiian Tree Snails," *Invertebrate Biology* 137, no. 2 (2018): 8.

41   Douglas K. Candland, "The Animal Mind and Conservation of Species: Knowing What Animals Know," *Current Science* 89, no. 7 (2005); Oded Berger-Tal et al., "Integrating Animal Behavior and Conservation Biology: A Conceptual Framework," *Behavioral Ecology* 22 (2011).

42   Val Plumwood, *Environmental Culture: The Ecological Crisis of Reason* (London: Routledge, 2002), 132. Of course, this "human" way of perceiving and knowing is also profoundly

shaped by one's culture, gender, language, and more. As decades of scholarship have shown, even at the most "basic" sensory level, there is no singular human experience of the world. See, for example, David Howes, ed., *Empire of the Senses: The Sensual Culture Reader* (Oxford: Berg Publishers, 2004).

43   E. Hayward, "Sensational Jellyfish: Aquarium Affects and the Matter of Immersion," *differences* 23, no. 3 (2012): 177.

44   Vinciane Despret, "It Is an Entire World That Has Disappeared," in *Extinction Studies: Stories of Time, Death and Generations*, ed. Deborah Bird Rose, Thom van Dooren, and Matthew Chrulew (New York: Columbia University Press, 2017).

45   Eileen Crist, "Ecocide and the Extinction of Animal Minds," in *Ignoring Nature No More: The Case for Compassionate Conservation*, ed. Marc Bekoff (Chicago: University of Chicago Press, 2013), 59.

CHAPTER 2

1   Brenden S. Holland, Luciano M. Chiaverano, and Cierra K. Howard, "Diminished Fitness in an Endemic Hawaiian Snail in Nonnative Host Plants," *Ethology Ecology & Evolution* 29, no. 3 (2017).

2   For a discussion of the history of biogeography as a field of scientific enquiry, see David Quammen, *The Song of the Dodo: Island Biogeography in an Age of Extinctions* (New York: Scribner, 1996).

3   Robert Macfarlane, *Underland: A Deep Time Journey* (London: Penguin, 2019).

4   Brendan S. Holland and Robert H. Cowie, "A Geographic Mosaic of Passive Dispersal: Population Structure in the Endemic Hawaiian Amber Snail *Succinea caduca* (Mighels, 1845)," *Molecular Ecology* 16, no. 12 (2007): 2432.

5   Brenden S. Holland, "Land Snails," in *Encyclopedia of Islands* (Los Angeles: University of California Press, 2009).

6   Małgorzata Ożgo et al., "Dispersal of Land Snails by Sea Storms," *Journal of Molluscan Studies* 82, no. 2 (2016).

7   Ożgo et al., "Dispersal of Land Snails by Sea Storms."

8   Ożgo et al., "Dispersal of Land Snails by Sea Storms."

9   W. J. Rees, "The Aerial Dispersal of Mollusca," *Journal of Molluscan Studies* 36, no. 5 (1965).

10   Dee S. Dundee, Paul H. Phillips, and John D. Newsom, "Snails on Migratory Birds," *Nautilus* 80 (1967): 90.

11   Joseph Vagvolgyi, "Body Size, Aerial Dispersal, and Origin of the Pacific Land Snail Fauna," *Systematic Biology* 24, no. 4 (1975).

12   Shinichiro Wada, Kazuto Kawakami, and Satoshi Chiba, "Snails Can Survive Passage through a Bird's Digestive System," *Journal of Biogeography* 39, no. 1 (2012).

13   This kind of dispersal and establishment of new species is often referred to as "colonization" by biologists. I avoid this term in this context, and have written elsewhere about the issues raised by this nomenclature in a colonized land: Thom van Dooren, "Moving Birds in Hawai'i: Assisted Colonisation in a Colonised Land," *Cultural Studies Review* 25, no. 1 (2019).

14   Sharon R. Kobayashi and Michael G. Hadfield, "An Experimental Study of Growth and Reproduction in the Hawaiian Tree Snails *Achatinella mustelina* and *Partulina redfieldii* (Achatinellinae)," *Pacific Science* 50, no. 4 (1996).

15   Robert H. Cowie and Brenden S. Holland, "Molecular Biogeography and Diversification of the Endemic Terrestrial Fauna of the Hawaiian Islands," *Philosophical Transactions B* 363, no. 1508 (2008). This number can be only an estimate for a variety of reasons, including the fact that some family groups may have arrived in the islands successfully on more than one occasion.

16   See Val Plumwood, "Nature in the Active Voice," *Australian Humanities Review* 46 (2009): 125.

17   Sato, Price, and Vaughan, "Kāhuli," 324–325. Thanks to Cody Pueo Pata for sharing the name "pūpūkaniao" with me in the form of a newspaper article from *Ka Nupepa Kuokoa*, "Haina Nane," June 4, 1909 (vol. 46, no. 23).

18   Yoshio Kondo, "Whistling Land Snails. Letter to Dr. Roland W. Force," Bishop Museum Library manuscript (1965).

19   Sato, Price, and Vaughan, "Kāhuli," 325.

20   Sato, Price, and Vaughan, "Kāhuli," 325.

21   Sato, Price, and Vaughan, "Kāhuli," 325.

22   Holland, Chiaverano, and Howard, "Diminished Fitness in an Endemic Hawaiian Snail in Nonnative Host Plants."

23   Rob Nixon, *Slow Violence and the Environmentalism of the Poor* (Cambridge, MA: Harvard University Press, 2011).

24   Michael G. Hadfield, "Extinction in Hawaiian Achatinelline Snails," *Malacologia* 27, no. 1 (1986). Also see C. Régnier et al., "Extinction in a Hyperdiverse Endemic Hawaiian Land Snail Family and Implications for the Underestimation of Invertebrate Extinction.," *Conservation Biology* 29, no. 6 (2015).

25   Terry L. Hunt, "Rethinking Easter Island's Ecological Catastrophe," *Journal of Archaeological Science* 34, no. 3 (2007); J. Stephen Athens, "*Rattus exulans* and the Catastrophic Disappearance of Hawai'i's Native Lowland Forest," *Biological Invasions* 11, no. 7 (2009).

26   Gon and Winter, "A Hawaiian Renaissance."

27   Paul D'Arcy, *Transforming Hawai'i: Balancing Coercion and Consent in Eighteenth-Century Kānaka Maoli Statecraft* (Canberra: ANU Press, 2018), 208; Carol A. MacLennan,

Sovereign Sugar: Industry and Environment in Hawai'i (Honolulu: University of Hawai'i Press, 2014), 23, 27.

28   It is unclear what form this protection took. Vancouver seems to have requested that a 10-year *kapu* (or taboo) be placed on their killing to ensure that they weren't all killed before having time to establish a breeding population in the islands. Whether this particular form of protection was afforded to them is debated. Nonetheless, as historian Jennifer Newell has put it, they did become "chiefly cattle," "closely enough allied with Kamehameha for them to be protected." Jennifer Newell, *Trading Nature: Tahitians, Europeans and Ecological Exchange* (Honolulu: University of Hawai'i Press, 2010), 135.

29   Newell, *Trading Nature*, 136.

30   Referenced in John Ryan Fischer, "Cattle in Hawai'i: Biological and Cultural Exchange," *Pacific Historical Review* 76, no. 3 (2007): 359.

31   Patricia Tummons, "First the Cattle, then the Bombs, Oust Hawaiians from Makua Valley," *Environment Hawai'i* 3, no. 5 (1992).

32   Deborah Bird Rose, *Reports from a Wild Country: Ethics for Decolonisation* (Sydney: UNSW Press, 2004), 85.

33   Seth Archer, *Sharks upon the Land: Colonialism, Indigenous Health, and Culture in Hawai'i, 1778–1855* (Cambridge: Cambridge University Press, 2018), 2.

34   MacLennan, *Sovereign Sugar*, 3.

35   MacLennan, *Sovereign Sugar*, 29.

36   MacLennan, *Sovereign Sugar*, 166–169.

37   Patricia Tummons, "Terrestrial Ecosystems," in *The Value of Hawai'i: Knowing the Past, Shaping the Future*, ed. Craig Howes and Jonathan K. K. Osorio (Honolulu: University of Hawai'i Press, 2010), 164–165.

38   Territory of Hawaii, *Report of the Commissioner of Agriculture and Forestry 1902* (Honolulu: Gazette Press, 1903).

39   Donna Haraway, "Anthropocene, Capitalocene, Plantationocene, Chthulucene: Making Kin," *Environmental Humanities* 6 (2015).

40   R. O'Rorke et al., "Not Just Browsing: An Animal That Grazes Phyllosphere Microbes Facilitates Community Heterogeneity," *ISME Journal* 11, no. 8 (2017); R. O'Rorke et al., "Dining Local: The Microbial Diet of a Snail That Grazes Microbial Communities Is Geographically Structured," *Environmental Microbiology* 17, no. 5 (2015).

41   Wallace M. Meyer, Rebecca Ostertag, and Robert H. Cowie, "Influence of Terrestrial Molluscs on Litter Decomposition and Nutrient Release in a Hawaiian Rain Forest," *Biotropica* 45, no. 6 (2013).

42   Storrs L. Olson and Helen F. James, "Descriptions of Thirty-Two New Species of Birds from the Hawaiian Islands: Part I. Non-Passeriformes," *Ornithological Monographs* 45 (1991): 25.

43   Anonymous, "One Last Effort to Save the Poʻouli" (Honolulu: Hawaiʻi Department of Land and Natural Resources, US Fish and Wildlife Service—Pacific Islands, and Zoological Society of San Diego, 2003).

44   On the history of ecology and biodiversity conservation in the United States, see Donald Worster, *Nature's Economy: A History of Ecological Ideas* (Cambridge: Cambridge University Press, 1994); David Takacs, *The Idea of Biodiversity: Philosophies of Paradise* (Baltimore: Johns Hopkins University Press, 1996).

45   Aldo Leopold, *A Sand County Almanac, and Sketches Here and There* (New York: Oxford University Press, 1949).

46   Sherwin Carlquist in Puanani O. Anderson-Fung and Kepā Maly, "Hawaiian Ecosystems and Culture: Why Growing Plants for Lei Helps to Preserve Hawaiʻi's Natural and Cultural Heritage," in *Growing Plants for Hawaiian Lei: 85 Plants for Gardens, Conservation, and Business*, ed. James R. Hollyer (Honolulu: College of Tropical Agriculture and Human Resources, University of Hawaiʻi, 2017).

47   Brenden S. Holland, and Michael G. Hadfield, "Islands within an Island: Phylogeography and Conservation Genetics of the Endangered Hawaiian Tree Snail *Achatinella mustelina*," *Molecular Ecology* 11, no. 3 (2002).

48   These are, respectively, examples of processes that biogeographers refer to as "dispersal" and "vicariance."

49   Baldwin, "The Land Shells of the Hawaiian Islands."

50   Richard Lewontin, *The Triple Helix: Gene, Organism, and Environment* (Cambridge, MA: Harvard University Press, 2000); Susan Oyama, *Evolution's Eye: A Systems View of the Biology–Culture Divide* (Durham, NC: Duke University Press, 2000).

51   Tim Ingold, "An Anthropologist Looks at Biology," *Man* (n.s.) 25, no. 2 (1990); Dorian Sagan, "Samuel Butler's Willful Machines," *The Common Review* 9 (2011).

52   Karola Stotz, "Extended Evolutionary Psychology: The Importance of Transgenerational Developmental Plasticity," *Frontiers in Psychology* 5, no. 908 (2014).

53   Ron Amundson, "John T. Gulick and the Active Organism: Adaptation, Isolation, and the Politics of Evolution," in *Darwin's Laboratory: Evolutionary Theory and Natural History in the Pacific*, ed. Roy M. MacLeod and Philip F. Rehbock (Honolulu: University of Hawaiʻi Press, 1994); Rebecca J. Rundell, "Snails on an Evolutionary Tree: Gulick, Speciation, and Isolation," *American Malacological Bulletin* 29, nos. 1–2 (2011).

54   Fascinatingly, Gulick was also an early proponent of the understanding that organisms might shape their own evolutionary trajectory. As Ron Amundson has noted: "At least ten years before the Baldwin Effect was proposed, Gulick had pointed out that when an organism behaves in a new way, it places itself in new relations to the external environment, and thereby changes the selective forces which influence the evolution of its lineage." Amundson, "John T. Gulick and the Active Organism," 132.

55 Epeli Hauʻofa, "Our Sea of Islands," in *A New Oceania: Rediscovering Our Sea of Islands*, ed. Eric Waddell, Vijay Naidu, and Epeli Hauʻofa (Suva, Fiji: University of the South Pacific, 1993), 2–16; 7.

56 Tracey Banivanua Mar, *Decolonisation and the Pacific: Indigenous Globalisation and the Ends of Empire* (Cambridge: Cambridge University Press, 2016), 17.

57 Hauʻofa, "Our Sea of Islands," 4. Thank you to Katerina Teaiwa and the other participants at BLUE Openings: Approaches to More-Than-Human Oceans for a helpful discussion on the connections between these literatures and biogeography (November 16–17, 2021, Aarhus University, Denmark).

58 Lawrence R. Heaney, "Is a New Paradigm Emerging for Oceanic Island Biogeography?," *Journal of Biogeography* 34 (2007): 753–757.

59 Elizabeth M. DeLoughrey, "The Myth of Isolates: Ecosystem Ecologies in the Nuclear Pacific," *Cultural Geographies* 20, no. 2 (2013): 167–184.

60 For further information on this project, see www.manoacliffreforestation.wordpress.com.

61 Holland, Chiaverano, and Howard, "Diminished Fitness," 237.

62 David E. Cooper, *The Measure of Things: Humanism, Humility, and Mystery* (Oxford: Oxford University Press, 2007).

63 Deborah Bird Rose, "Pattern, Connection, Desire: In Honour of Gregory Bateson," *Australian Humanities Review* 35 (2005).

64 Noelani Arista, "Navigating Uncharted Oceans of Meaning: Kaona as Historical and Interpretive Method," *PMLA* 125, no. 3 (2010): 666.

65 This is a topic that I have explored in greater detail in van Dooren, *Flight Ways*, 21–44.

CHAPTER 3

1 Vernadette Vicuna Gonzalez, *Securing Paradise: Tourism and Militarism in Hawaiʻi and the Philippines* (Durham, NC: Duke University Press Books, 2013), 67–81; Eileen Momilani Naughton, "The Bernice Pauahi Bishop Museum: A Case Study Analysis of Mana as a Form of Spiritual Communication in the Museum Setting" (PhD diss., Simon Fraser University, 2001), 149–171.

2 USFWS, "Endangered and Threatened Wildlife and Plants," 46 FR 3178.

3 https://www.reaganlibrary.gov/archives/speech/remarks-announcing-establishment-presidential-task-force-regulatory-relief. This announcement was followed up by the issuing of Executive Order 12291, available at https://www.archives.gov/federal-register/codification/executive-order/12291.html, accessed May 25, 2021.

4 USFWS, "Endangered and Threatened Wildlife and Plants; Deferral of Effective Dates," *Federal Register* 46 FR 40025 (1981).

5   Zygmunt J. B. Plater, "In the Wake of the Snail Darter: An Environmental Law Paradigm and Its Consequences," *Journal of Law Reform* 19, no. 4 (1986): 829.

6   Wayne C. Gagné, "Conservation Priorities in Hawaiian Natural Systems," *BioScience* 38, no. 4 (1988): 268.

7   http://papahanakuaola.com, accessed November 16, 2018.

8   Candace Kaleimamoowahinekapu Galla et al., "Perpetuating Hula: Globalization and the Traditional Art," *Pacific Arts Association* 14, no. 1 (2015).

9   The term, as Anderson-Fung and Maly explain, "is derived from the words kino, meaning 'form or embodiment,' and lau, meaning 'many.' Some believe that virtually every plant species known to the Hawaiians was considered kinolau of some spirit or deity." Anderson-Fung and Maly, "Hawaiian Ecosystems and Culture," 13.

10 Anderson-Fung and Maly, "Hawaiian Ecosystems and Culture"; Chai Kaiaka Blair-Stahn, "The Hula Dancer as Environmentalist: (Re-)Indigenising Sustainability through a Holistic Perspective on the Role of Nature in Hula," in *Proceedings of the 4th International Traditional Knowledge Conference*, ed. Joseph S. Te Rito and Susan M. Healy, 60–64 (Auckland: Knowledge Exchange Programme of Ngā Pae o te Māramatanga, 2010).

11 Cited in Blair-Stahn, "The Hula Dancer as Environmentalist."

12 For detailed discussions of the history and contemporary reality of colonization in Hawai'i, see Haunani-Kay Trask, *From a Native Daughter: Colonialism and Sovereignty in Hawai'i* (Honolulu: University of Hawai'i Press, 1999); Osorio, *Dismembering Lāhui*; Noenoe K. Silva, *Aloha Betrayed: Native Hawaiian Resistance to American Colonialism* (Durham, NC: Duke University Press, 2004); Noenoe K. Silva, *The Power of the Steel-Tipped Pen: Reconstructing Native Hawaiian Intellectual History* (Durham, NC: Duke University Press, 2017); Archer, *Sharks upon the Land*; Noelani Goodyear-Ka'ōpua, Ikaika Hussey, and Erin Kahunawaika'ala, eds., *A Nation Rising: Hawaiian Movements for Life, Land, and Sovereignty* (Durham, NC: Duke University Press, 2014).

13 Yoshio Kondo and William J. Clench, "Charles Montague Cooke, Jr.: A Bio-Bibliography," *Bernice P. Bishop Museum, Special Publication* 42 (1952): 9.

14 The characterization of Hawai'i as an "occupied" or "colonized" land is complex and contentious, with each term invoking a somewhat different understanding of the past and current situation while also framing the kinds of legal and political responses that might be taken up in struggles for sovereignty. On this topic, see J. Kēhaulani Kauanui, *Paradoxes of Hawaiian Sovereignty: Land, Sex, and the Colonial Politics of State Nationalism* (Durham, NC: Duke University Press, 2018).

15 Leon No'eau Peralto, "O Koholālele, He 'Āina, He Kanaka, He I'a Nui Nona Ka Lā: Re-Membering Knowledge of Place in Koholālele, Hāmākua, Hawai'i," in *I Ulu I Ka 'Āina: Land*, ed. Jonathan Osorio (Honolulu: University of Hawai'i Press, 2014), 76.

16 Sato, Price, and Vaughan, "Kāhuli," 330.

17   Charles Samuel Stewart, "Addenda: *Achatina stewartii* and *A. oahuensis* [1823]," *Manual of Conchology* 22 (1912).

18   Beckwith, *The Kumulipo: A Hawaiian Creation Chant*.

19   Anderson-Fung and Maly, "Hawaiian Ecosystems and Culture."

20   https://www.hawaiipublicradio.org/post/episode-3-meaning-aloha-ina-puanani-burgess, accessed May 25, 2021.

21   Jonathan Goldberg-Hiller and Noenoe K. Silva, "Sharks and Pigs: Animating Hawaiian Sovereignty against the Anthropological Machine," *The South Atlantic Quarterly* 110 (2011): 436.

22   Goldberg-Hiller and Silva, "Sharks and Pigs," 436.

23   Sato, Price, and Vaughan, "Kāhuli," 324–325.

24   https://manomano.io/, accessed May 25, 2021.

25   Noah Gomes, "Reclaiming Native Hawaiian Knowledge Represented in Bird Taxonomies," *Ethnobiology Letters* 11, no. 2 (2020).

26   Gomes, "Reclaiming Native Hawaiian Knowledge," 34.

27   Gomes, "Reclaiming Native Hawaiian Knowledge."

28   Hadfield, "Extinction in Hawaiian Achatinelline Snails."

29   Baldwin, "The Land Shells of the Hawaiian Islands." 55

30   Osorio, *Dismembering Lāhui*, 9–13.

31   Robert E. Kohler, "Finders, Keepers: Collecting Sciences and Collecting Practice," *History of Science* 45, no. 4 (2007): 438.

32   E. Alison Kay, "Missionary Contributions to Hawaiian Natural History: What Darwin Didn't Know," *The Hawaiian Journal of History* 31 (1997).

33   Addison Gulick, *Evolutionist and Missionary, John Thomas Gulick: Portrayed through Documents and Discussions* (Chicago: University of Chicago Press, 1932), 112.

34   T. C. B. Rooke, "The Sandwich Island Institute: Inaugural Thesis, Delivered before the Sandwich Island Institute, Dec. 12, 1838." *The Hawaiian Spectator*, 1838.

35   Baldwin, "The Land Shells of the Hawaiian Islands," 55–56.

36   "Wedding Bells," *The Daily Bulletin*, Honolulu, HI, January 2, 1885, 1.

37   Kay, "Missionary Contributions," 39.

38   Hadfield, "Extinction in Hawaiian Achatinelline Snails," 70.

39   Jonathan Galka, "Mollusk Loves: Becoming with Native and Introduced Land Snails in the Hawaiian Islands," *Island Studies Journal* (forthcoming 2021).

40   Gulick, *Evolutionist and Missionary*, 120–121.

41   Archer, *Sharks upon the Land*, 225, 59–60.

42   Archer, *Sharks upon the Land*, 226.

43   Gulick, *Evolutionist and Missionary*, 125–126.

44   Gulick, *Evolutionist and Missionary*, 125–126.

45   Osorio, *Dismembering Lāhui*; Stuart Banner, *Possessing the Pacific: Land, Settlers, and Indigenous People from Australia to Alaska* (Cambridge, MA: Harvard University Press, 2007); Silva, *Aloha Betrayed*.

46   Davianna Pomaikaʻi McGregor, "Waipiʻo Valley, a Cultural *Kīpuka* in Early 20th Century Hawaiʻi," *The Journal of Pacific History* 30, no. 2 (1995): 194–195.

47   MacLennan, *Sovereign Sugar*.

48   J. Kēhaulani Kauanui, *Hawaiian Blood: Colonialism and the Politics of Sovereignty and Indigeneity* (Durham, NC: Duke University Press, 2008), 75.

49   Puakea Nogelmeier, "Mai Paʻa I Ka Leo: Historical Voice in Hawaiian Primary Materials, Looking Forward and Listening Back" (PhD diss., University of Hawaiʻi at Mānoa, 2003), xii.

50   kuʻualoha hoʻomanawanui, *Voices of Fire: Reweaving the Literary Lei of Pele and Hiʻiaka* (Minneapolis: University of Minnesota Press, 2014), 33–64; Noelani Arista, "Listening to Leoiki: Engaging Sources in Hawaiian History," *Biography* 32, no. 1 (2009); Silva, *The Power of the Steel-Tipped Pen*.

51   Silva, *Aloha Betrayed*.

52   H. M. Ayres, "Under the Coconut Tree," *The Hawaiian Star*, December 16, 1911.

53   MacLeod and Rehbock, *Darwin's Laboratory*; Tom Griffiths and Libby Robin, eds., *Ecology and Empire: Environmental History of Settler Societies* (Seattle: University of Washington Press, 1997); Paula Findlen, *Possessing Nature: Museums, Collecting, and Scientific Culture in Early Modern Italy* (Berkeley: University of California Press, 1994); Nicholas Jardine, James A.Secord, and Emma C. Spary, eds., *Cultures of Natural History* (Cambridge: Cambridge University Press, 1996).

54   Libby Robin, "Ecology: A Science of Empire," in *Ecology and Empire: Environmental History of Settler Societies*, ed. Tom Griffiths and Libby Robin (South Carlton: Melbourne University Press, 1997).

55   MacLennan, *Sovereign Sugar*.

56   C. M. Hyde, "Exhibition of Land Shells," *The Pacific Commercial Advertiser*, February 28, 1895, 1.

57   Sune Borkfelt, "What's in a Name?: Consequences of Naming Non-Human Animals," *Animals* 1 (2011); Etienne S. Benson, "Naming the Ethological Subject," *Science in Context* 29, no. 1 (2016).

58   Russell Clement, "From Cook to the 1840 Constitution: The Name Change from Sandwich to Hawaiian Islands," *Hawaiian Journal of History* 14 (1980).

59 Kay, "Missionary Contributions," 39.

60 Anonymous, "Mount Tantalus Got Its Name from Punahou Boys," *The Pacific Commercial Advertiser*, October 7, 1901.

61 McDougall, *Finding Meaning*.

62 hoʻomanawanui, *Voices of Fire*, 55.

63 Writing in the context of the colonization of Aotearoa/New Zealand, Maori scholar Linda Tuhiwai Smith has argued: "Renaming the land was probably as powerful ideologically as changing the land. Indigenous children in schools, for example, were taught the new names for places that they and their parents had lived in for generations. These were the names which appeared on maps and which were used in official communications. This newly named land became increasingly disconnected from the songs and chants used by indigenous peoples to trace their histories, to bring forth spiritual elements or to carry out the simplest of ceremonies." Linda Tuhiwai Smith, *Decolonizing Methodologies: Research and Indigenous Peoples* (London: Zed Books, 2012), 107.

64 David D. Baldwin, "Descriptions of New Species of Achatinellidae from the Hawaiian Islands," *Proceedings of the Academy of Natural Sciences of Philadelphia* 47 (1895).

65 The genus *Achatinella* was a popular one among the snails for these kinds of bestowals. In addition to *Achatinella dolei*, these species include *A. juddii*, *A. byronii*, *A. lyonsiana*, *A. cookei*, *A. stewartii*, and *A. spaldingi*. Many of these species were subsequently "revised" and so are no longer officially recognized (discussed further in chapter 4). The others are all either extinct or critically endangered.

66 Baldwin, "Descriptions of New Species," 220.

67 W. R. Farrington, "Editorial," *The Pacific Commercial Advertiser*, March 1, 1895, 1.

68 Hadfield, "Extinction in Hawaiian Achatinelline Snails," 73.

69 Hadfield, "Extinction in Hawaiian Achatinelline Snails," 74.

70 Hadfield, "Extinction in Hawaiian Achatinelline Snails," 74.

71 See, for example, Baldwin, "The Land Shells of the Hawaiian Islands"; Gulick, *Evolutionist and Missionary*, 411.

72 Anonymous, "Notes of the Week," *The Hawaiian Gazette*, October 8, 1873.

73 Sepkoski, *Catastrophic Thinking*. On this topic, also see Patrick Brantlinger, *Dark Vanishings: Discourse on the Extinction of Primitive Races, 1800–1930* (Ithaca, NY: Cornell University Press, 2003); Miles A. Powell, *Vanishing America: Species Extinction, Racial Peril, and the Origins of Conservation* (Cambridge, MA: Harvard University Press, 2016).

74 Joseph J. Gouveia, "A Collecting Trip on the Island of Oahu, Hawaiian Islands, by the Gulick Natural History Club," *Nautilus*, 33, no. 2 (1919): 54–58.

75 David D. Baldwin, "Land Shell Collecting on Oahu," *Hawaii's Young People* 4, no. 8 (1900): 240.

76 This tendency has a long and ongoing history and includes the collection/documentation of both human and nonhuman forms of diversity. For historical discussions, see T. Griffiths, *Hunters and Collectors: The Antiquarian Imagination in Australia* (Cambridge: Cambridge University Press, 1996); and Robert E. Kohler, *All Creatures: Naturalists, Collectors, and Biodiversity, 1850–1950* (Princeton, NJ: Princeton University Press, 2013). For contemporary discussions, see Jenny Reardon, *Race to the Finish: Identity and Governance in an Age of Genomics* (Princeton, NJ: Princeton University Press, 2005); Carrie Friese, *Cloning Wild Life: Zoos, Captivity, and the Future of Endangered Animals* (New York: NYU Press, 2013); and Joanna Radin and Emma Kowal, eds., *Cryopolitics: Frozen Life in a Melting World* (Cambridge, MA: MIT Press, 2017).

77 Haunani-Kay Trask, "The Birth of the Modern Hawaiian Movement: Kalama Valley, Oʻahu," *Hawaiian Journal of History* 21 (1987); Goodyear-Kaʻōpua, Hussey, and Kahunawaikaʻala, *A Nation Rising*.

78 Mehana Blaich Vaughan, *Kaiaulu: Gathering Tides* (Corvallis: Oregon State University Press, 2018); Gon and Winter, "A Hawaiian Renaissance"; Candace Fujikane, *Mapping Abundance for a Planetary Future: Kanaka Maoli and Critical Settler Cartographies in Hawaiʻi* (Durham, NC: Duke University Press, 2021); Kawika Winter et al., "The Moku System: Managing Biocultural Resources for Abundance within Social-Ecological Regions in Hawaiʻi," *Sustainability* 10 (2018); Noelani Goodyear-Kaʻōpua, "Rebuilding the ʻAuwai: Connecting Ecology, Economy and Education in Hawaiian Schools," *AlterNative: An International Journal of Indigenous Peoples* 5, no. 2 (2009).

79 Kali Fermantez, "Re-Placing Hawaiians in Dis Place We Call Home," *Hūlili: Multidisciplinary Research on Hawaiian Well-Being* 8 (2012).

80 Gerald Vizenor, *Survivance: Narratives of Native Presence* (Lincoln: University of Nebraska Press, 2008); Brandy Nālani McDougall and Georganne Nordstrom, "Ma Ka Hana Ka ʻIke (In the Work Is the Knowledge): Kaona as Rhetorical Action," *College Composition and Communication* 63, no. 1 (2011).

81 McGregor, "Waipiʻo Valley." I have used the term "(re)connection" here to describe this process. In doing so I am guided by Cutcha Risling Baldy, a Hupa, Yurok, and Karuk scholar who, drawing on the work of Mishuana Goeman, has argued that this "(re)" prefix ought to be understood as connoting the fact that "Indigenous peoples are not just claiming and naming (or mapping, or creating) in the present but they are participating in a (re)vitalization that builds a future with the past, and shows how these epistemological foundations speak to a lasting legacy, that is both ancient and modern in their discourse that challenges settler colonialism." Cutcha Risling Baldy, "Coyote Is Not a Metaphor: On Decolonizing, (Re)claiming and (Re)naming ʻCoyote,'" *Decolonization: Indigeneity, Education & Society* 4, no. 1 (2015).

82 Amy Kuʻuleialoha Stillman, "The Hawaiian *Hula* and Legacies of Institutionalization," *Comparative American Studies: An International Journal* 5, no. 2 (2007); Nogelmeier, "Mai Paʻa I Ka Leo"; Arista, "Listening to Leoiki."

83 For a detailed discussion of the history and politics of taxonomic naming, see Stephen B. Heard, *Charles Darwin's Barnacle and David Bowie's Spider: How Scientific Names Celebrate Adventurers, Heroes, and Even a Few Scoundrels* (New Haven, CT: Yale University Press, 2020).

84 Kate Evans, "Change Species Names to Honor Indigenous Peoples, Not Colonizers, Researchers Say," *Scientific American* (2020), https://www.scientificamerican.com/article /change-species-names-to-honor-indigenous-peoples-not-colonizers-researchers-say/.

85 As Stephen Heard has noted: "Expeditions and collecting trips often had Indigenous guides, field assistants, and other support workers, and the contributions these people made weren't trivial—many an expedition would have failed miserably without them. A recent compilation by John van Wyhe, for example, suggests that Alfred Russel Wallace's famous expedition to the Malay Archipelago probably involved well over 1,000 local assistants." Heard, *Charles Darwin's Barnacle.*

86 Silva, *The Power of the Steel-Tipped Pen;* Arista, "Listening to Leoiki"; Nogelmeier, "Mai Pa'a I Ka Leo."

87 Sato, Price, and Vaughan, "Kāhuli," 330.

CHAPTER 4

1 José Antonio González-Oreja, "The Encyclopedia of Life vs. the Brochure of Life: Exploring the Relationships between the Extinction of Species and the Inventory of Life on Earth," *Zootaxa* 1965, no. 1 (2008).

2 C. Mora, A. Rollo, and D. P. Tittensor, "Comment on 'Can We Name Earth's Species before They Go Extinct?'" *Science* 341, no. 6143 (2013); González-Oreja, "The Encyclopedia of Life."

3 Andy Purvis, "A Million Threatened Species? Thirteen Questions and Answers" (2019): https://ipbes.net/news/million-threatened-species-thirteen-questions-answers, accessed May 25, 2021.

4 Kondo and Clench, "Charles Montague Cooke, Jr.," 4–5. Prior to taking up this position, Cooke was an assistant in the collection from 1902.

5 Robert H. Cowie, "Yoshio Kondo: Bibliography and List of Taxa," *Bishop Museum Occasional Papers* 32 (1993); Carl C. Christensen, "Dr. Kondo Retires as Bishop Museum Malacologist," *Hawaiian Shell News* 29, no. 1 (1981).

6 It bears noting that the pineapple was one of the principal plantation crops of Hawai'i, a plant thoroughly implicated in the destruction of the environment and the displacement of Kānaka Maoli from their ancestral lands. It is also a crop that Cooke's family had benefited significantly from financially. Cooke's grandfather was Amos Starr Cooke, who, among his many other important roles in the kingdom and then territory of Hawai'i, cofounded Castle & Cooke, one of the "Big Five" companies that wielded considerable political power in the islands. Castle

& Cooke was a major pineapple producer, including through its subsidiary Dole, a company whose name is still synonymous with the pineapple today. Thus, the alcohol that preserved these snail bodies was in important ways inseparable from the larger story of colonization, wealth, and privilege that played no small role in enabling Cooke's important work.

7   Kondo and Clench, "Charles Montague Cooke, Jr."

8   Christensen, "Dr. Kondo Retires."

9   William J. Clench, "John T. Gulick's Hawaiian Land Shells," *Nautilus* 72 (1959).

10 Yeung and Hayes, "Biodiversity and Extinction of Hawaiian Land Snails," 1160. Thank you to Nori for additional clarifications and estimates of the collection's past and current holdings, quite a few of which, as we will see, are not yet fully catalogued.

11 https://conchologistsofamerica.org/bynes-disease-questions-and-answers/, accessed May 25, 2021.

12 Bill Wood, "Hard Choices at Bishop Museum," *Hawaii Investor* (October 1985); Naughton, "The Bernice Pauahi Bishop Museum," 172–187.

13 Robert H. Cowie, Neal L. Evenhuis, and Carl C. Christensen, *Catalog of the Native Land and Freshwater Molluscs of the Hawaiian Islands* (Leiden: Backhuys Publishers, 1995).

14 In creating this catalogue, Cowie and his colleagues compiled a list of all recognized Hawaiian land snail species, indicating a total number of 752. Since its publication, however, seven additional species have been added to that list. Two new species, *Partulina puupiliensis* and *Partulina hobdyi*, were described in Mike Severns, *Shells of the Hawaiian Islands: The Land Shells* (Hackenheim: ConchBooks, 2011). In addition, Severns elevated three Big Island subspecies to full species. Two other new species, *Auriculella gagneorum* and *Endodonta christenseni*, have since been described by Nori, Ken, and their colleagues: Kenneth A. Hayes et al., "The Last Known *Endodonta* Species? *Endodonta christenseni* sp. nov. (Gastropoda: Endodontidae)," *Bishop Museum Occasional Papers* 138 (2020); Norine W. Yeung et al., "Overlooked but Not Forgotten: The First New Extant Species of Hawaiian Land Snail Described in 60 Years, *Auriculella gagneorum* sp. nov. (Achatinellidae, Auriculellinae)," *ZooKeys* 950 (2020). Thank you to Rob Cowie for this extended explanation of additional species described since his 1995 catalogue.

15 Wayne C. Gagne and Carl C. Christensen, "Conservation Status of Native Terrestrial Invertebrates in Hawai'i," in *Hawaii's Terrestrial Ecosystems: Preservation and Management*, ed. C. P. Stone and J. M. Scott (Honolulu: Cooperative National Park Resources Studies Unit, University of Hawai'i, 1985).

16 Yeung and Hayes, "Biodiversity and Extinction of Hawaiian Land Snails," 1159.

17 Figures shared at Hui Kāhuli meeting organized by Dave Sischo, Bishop Museum, Honolulu, HI, March 9, 2020.

18 C. Régnier et al., "Foot Mucus Stored on FTA® Cards Is a Reliable and Non-Invasive Source of DNA for Genetics Studies in Molluscs," *Conservation Genetics Resources* 3, no. 2 (2011).

19 Baldwin, "The Land Shells of the Hawaiian Islands," 57.

20 Robert Cameron, *Slugs and Snails*, 19.

21 Kondo and Clench, "Charles Montague Cooke, Jr.," 9.

22 Alain Dubois, "Describing New Species," *Taprobanica: The Journal of Asian Biodiversity* 2, no. 1 (2011): 7.

23 For a fascinating discussion of *Liguus* snails and the people who study them, see Jonathan Galka, "*Liguus* Landscapes: Professional Malacology, Amateur Ligging, and the Social Life of Snail Science," *Journal of the History of Biology* 55, no. 4 (2022).

24 Lorraine Daston, "Type Specimens and Scientific Memory," *Critical Inquiry* 31 (2004): 156.

25 Joshua Trey Barnett, "Naming, Mourning, and the Work of Earthly Coexistence," *Environmental Communication* 13, no. 3 (2019): 294.

26 Prather et al., "Invertebrates, Ecosystem Services and Climate Change"; Nico Eisenhauer, Aletta Bonn, and Carlos A. Guerra, "Recognizing the Quiet Extinction of Invertebrates," *Nature Communications* 10, no. 1 (2019); Francisco Sánchez-Bayo and Kris A. G. Wyckhuys, "Worldwide Decline of the Entomofauna: A Review of Its Drivers," *Biological Conservation* 232 (2019).

27 Cowie et al., "Measuring the Sixth Extinction."

28 J. E. Baillie, C. Hilton-Taylor, and S. N. Stuart, eds., *A Global Species Assessment* (Gland: IUCN, 2004); Régnier, Fontaine, and Bouchet, "Not Knowing, Not Recording, Not Listing."

29 S. N. Stuart et al., "The Barometer of Life," *Science* 328, no. 5975 (2010).

30 Michael R. Donaldson et al., "Taxonomic Bias and International Biodiversity Conservation Research," *FACETS* 1 (2016).

31 Cowie et al., "Measuring the Sixth Extinction," 4.

32 For the authoritative, frequently updated database of mollusk species, see MolluscaBase: https://www.molluscabase.org/

33 Eisenhauer, Bonn, and Guerra, "Recognizing the Quiet Extinction of Invertebrates."

34 A. Dubois, "The Relationships between Taxonomy and Conservation Biology in the Century of Extinctions," *Comptes Rendus Biologies* 326, Suppl. 1 (2003): S10.

35 González-Oreja, "The Encyclopedia of Life."

36 Benoît Fontaine, Adrien Perrard, and Philippe Bouchet, "21 Years of Shelf Life between Discovery and Description of New Species," *Current Biology* 22 (2012).

37 M. J. Costello, R. M. May, and N. E. Stork, "Can We Name Earth's Species before They Go Extinct?," *Science* 339, no. 6118 (2013).

38 A. F. Sartori, O. Gargominy, and B. Fontaine, "Anthropogenic Extinction of Pacific Land Snails: A Case Study of Rurutu, French Polynesia, with Description of Eight New Species of Endodontids (Pulmonata)," *Zootaxa* 3640 (2013): 343.

39 Yeung, and Hayes, "Biodiversity and Extinction of Hawaiian Land Snails," 1162.

40 Purvis, "A Million Threatened Species?"

41 Charles Lydeard et al., "The Global Decline of Nonmarine Mollusks," *BioScience* 54, no. 4 (2004). Figure for currently described species from https://www.molluscabase.org/, accessed May 20, 2021.

42 Lydeard et al., "The Global Decline of Nonmarine Mollusks."

43 Ira Richling and Philippe Bouchet, "Extinct even before Scientific Recognition: A Remarkable Radiation of Helicinid Snails (Helicinidae) on the Gambier Islands, French Polynesia," *Biodiversity and Conservation* 22, no. 11 (2013).

44 Sartori, Gargominy, and Fontaine, "Anthropogenic Extinction of Pacific Land Snails."

45 Michelle Bastian, "Whale Falls, Suspended Ground, and Extinctions Never Known," *Environmental Humanities* 12, no. 2 (2020).

46 Cameron, *Slugs and Snails*, 38–48.

47 Dalia Nassar and Margaret M. Barbour, "Rooted," *Aeon* (October 16, 2019).

48 Richling and Bouchet, "Extinct even before Scientific Recognition."

49 Richling and Bouchet, "Extinct even before Scientific Recognition," 2442.

50 Cameron, *Slugs and Snails*, 330–32.

51 Melissa Mertl, "Taxonomy in Danger of Extinction," *Science Magazine* (May 22, 2002).

52 Richard Conniff, "Conservation Conundrum: Is Focusing on a Single Species a Good Strategy?," *Yale Environment 360* (May 17, 2018); Sandy J. Andelman and William F. Fagan, "Umbrellas and Flagships: Efficient Conservation Surrogates or Expensive Mistakes?," *Proceedings of the National Academy of Sciences* 97 (2000).

53 https://theecologist.org/2011/jan/02/species-vs-ecosystems-save-tiger-or-focus-bigger -issues, accessed May 25, 2021.

CHAPTER 5

1 Marion Kelly and Nancy Aleck, *Mākua Means Parents: A Brief Cultural History of Mākua Valley* (Honolulu: American Friends Service Committee, 1997), 2.

2 Tummons, "First the Cattle."

3 Patricia Tummons, "Army Tenure at Makua Valley Solidified after Statehood," *Environment Hawai'i* 3, no. 5 (1992).

4 Patricia Tummons, "Army's Application for EPA Permit Is Long, but Not Informative," *Environment Hawai'i* 3, no. 5 (1992).

5   Alison A. Dalsimer, "Threatened and Endangered Species on DOD Lands," *Department of Defense Natural Resources Program Fact Sheet* (2016).

6   Dalsimer, "Threatened and Endangered Species"; Bruce A. Stein, Cameron Scott, and Nancy Benton, "Federal Lands and Endangered Species: The Role of Military and Other Federal Lands in Sustaining Biodiversity," *BioScience* 58, no. 4 (2008).

7   US Department of Defense and US Fish & Wildlife Service, "The Military and the Endangered Species Act: Interagency Cooperation," *Factsheet* (2001).

8   https://www.hawaiibusiness.com/endangered-and-underfunded/, accessed May 25, 2021. Also see Leonard, "Recovery Expenditures."

9   For a detailed discussion of the history and future of struggles over culture, land, and sovereignty in Hawai'i, see Goodyear-Ka'ōpua, Hussey, and Kahunawaika'ala, *A Nation Rising*. See also Kauanui, *Paradoxes of Hawaiian Sovereignty*.

10   For discussions of tensions, collaborations, and conflicts between Indigenous peoples and conservationists, see Goldberg-Hiller and Silva, "Sharks and Pigs"; Thom van Dooren, *The Wake of Crows: Living and Dying in Shared Worlds* (New York: Columbia University Press, 2019), 71–94; Eve Vincent and Timothy Neale, eds., *Unstable Relations: Indigenous People and Environmentalism in Contemporary Australia* (Perth: UWA Publishing, 2016); Paige West, James Igoe and Dan Brockington, "Parks and Peoples: The Social Impact of Protected Areas," *Annual Review of Anthropology* 35 (2006); Anna Lowenhaupt Tsing, *Friction: An Ethnography of Global Connection* (Princeton, NJ: Princeton University Press, 2005); June Mary Rubis and Noah Theriault, "Concealing Protocols: Conservation, Indigenous Survivance, and the Dilemmas of Visibility," *Social & Cultural Geography* 21, no. 7 (2019); Alexander Mawyer and Jerry K Jacka, "Sovereignty, Conservation and Island Ecological Futures." *Environmental Conservation* 45, no. 3 (2018); Liv Østmo and John Law, "Mis/translation, Colonialism, and Environmental Conflict," *Environmental Humanities* 10, no. 2 (2018).

11   David Vine, *Base Nation: How US Military Bases abroad Harm America and the World* (New York: Henry Holt and Company, 2015).

12   Niall McCarthy, "Report: The U.S. Military Emits More CO2 than Many Industrialized Nations," *Forbes* (June 13, 2019).

13   Carl C. Christensen and Michael G. Hadfield, *Field Survey of Endangered O'ahu Tree Snails (Genus Achatinella) on the Makua Military Reservation* (Honolulu: Division of Malacology, Bernice P. Bishop Museum, 1984).

14   D'Alté A. Welch, "Distribution and Variation of *Achatinella mustelina* Mighels in the Waianae Mountains, Oahu," *Bernice P. Bishop Museum Bulletin* 152 (1938).

15   Christensen and Hadfield, *Field Survey*, 13.

16   Christensen and Hadfield, *Field Survey*, 16.

17   Patricia Tummons, "Endangered Snails of Makua Valley Are Placed at Risk by Army Fires," *Environment Hawai'i* 3, no. 5 (1992).

18 Tummons, "Endangered Snails of Makua Valley."

19 Kalamaokaʻāina Niheu, "Puʻuhonua: Sanctuary and Struggle at Mākua," in *A Nation Rising*, ed. Goodyear-Kaʻōpua, Hussey, and Kahunawaikaʻala.

20 Jonathan Kamakawiwoʻole Osorio, "Hawaiian Souls: The Movement to Stop the U.S. Military Bombing of Kahoʻolawe," in *A Nation Rising*, ed. Goodyear-Kaʻōpua, Hussey, and Kahunawaikaʻala.

21 https://earthjustice.org/news/press/2003/community-group-warns-army-to-reinitiate-fws -review-of-makua, accessed May 25, 2021.

22 The Makua Implementation Team, *Implementation Plan: Makua Military, Reservation, Island of Oahu* (Honolulu: United States Army Garrison, Hawaiʻi, 2003), 2–2.

23 Wording of the original permit cited in Kelly and Aleck, *Mākua Means Parents*, 9.

24 Tummons, "Army Tenure at Makua Valley Solidified after Statehood."

25 https://kahoolawe.hawaii.gov/history.shtml, accessed May 25, 2021.

26 David Havlick, "Logics of Change for Military-to-Wildlife Conversions in the United States," *GeoJournal* 69, no. 3 (2007); David G. Havlick, *Bombs Away: Militarization, Conservation, and Ecological Restoration* (Chicago: University of Chicago Press, 2018).

27 Havlick, "Logics of Change," 156.

28 Rick Zentelis and David Lindenmayer, "Bombing for Biodiversity-Enhancing Conservation Values of Military Training Areas," *Conservation Letters* 8, no. 4 (2015): 301.

29 Jobriath Rohrer et al., *Development of Tree Snail Protection Enclosures: From Design to Implementation* (Honolulu: Pacific Cooperative Studies Unit, University of Hawaiʻi at Mānoa, 2016).

30 Even this first Koʻolau exclosure was built by the Army, but it was then handed over to SEPP after it was determined that the military doesn't have a direct negative impact on the relevant species.

31 Maui Nui, "Greater Maui," is a name used mostly by geologists and biologists to refer collectively to the four islands of Maui, Molokaʻi, Lānaʻi, and Kahoʻolawe. These islands once formed a single larger island but have now been separated as a result of changes in sea level over the past million years.

CHAPTER 6

1 Thank you to Candace Fujikane for pointing out the appropriateness of the term "kīpuka" in this context.

2 Lesley Head, *Hope and Grief in the Anthropocene: Re-Conceptualising Human–Nature Relations* (London: Routledge, 2016).

3 Rebecca Solnit, *Hope in the Dark: Untold Histories, Wild Possibilities* (Edinburgh: Canongate Books, 2016), xii.

4   As I've noted in other work, "care" must be understood here as something far more than abstract well-wishing. To care for another, to care for a possible world, is to become emotionally and ethically entangled and consequently to get involved in whatever practical ways we can to help to bring it about. Van Dooren, *The Wake of Crows*.

5   A. F. James et al., "Modelling the Growth and Population Dynamics of the Exiled Stockton Coal Plateau Landsnail, *Powelliphanta augusta*," *New Zealand Journal of Zoology* 40, no. 3 (2013).

6   Eben Kirksey writes about bubbles of hope, happiness, and care in S. Eben Kirksey, *Emergent Ecologies* (Durham, NC: Duke University Press, 2015).

7   D. Brossard et al., "Promises and Perils of Gene Drives: Navigating the Communication of Complex, Post-Normal Science," *Proceedings of the National Academy of Sciences* 116, no. 16 (2019). Many of Hawai'i's endemic forest birds are now gone, and most of those that hang on do so in captive facilities or shrinking refugia. Of the 113 bird species known to have lived exclusively in the Hawaiian Islands just prior to human arrival, almost two-thirds are extinct. Of the 42 that remain, roughly three-quarters are federally listed under the ESA. See Leonard, "Recovery Expenditures for Birds." For those that do remain, restoration will likely require the eradication of introduced mosquitoes, vectors for devastating avian malaria. Here, too, climate change is exacerbating the situation, with warmer weather enabling mosquitoes to move into some of the few remaining higher-elevation populations of species like the 'akikiki (*Oreomystis bairdi*) and the 'akeke'e (*Loxops caeruleirostris*) on the island of Kaua'i.

8   Carl C. Christensen, "Should We Open This Can of Worms? A Call for Caution Regarding Use of the Nematode *Phasmarhabditis hermaphrodita* for Control of Pest Slugs and Snails in the United States," *Tentacle* 27 (2019).

9   J. Fischer and D. B. Lindenmayer, "An Assessment of the Published Results of Animal Relocations," *Biological Conservation* 96 (2000).

10   Head, *Hope and Grief*.

11   Anna Lowenhaupt Tsing, "Blasted Landscapes, and the Gentle Art of Mushroom Picking," in *The Multispecies Salon: Gleanings from a Para-Site*, ed. Eben Kirksey (Durham, NC: Duke University Press, 2016).

12   Chantal Mouffe and Ernesto Laclau, "Hope, Passion, Politics," in *Hope: New Philosophies for Change*, ed. Mary Zournarzi (Annandale, NSW: Pluto Press, 2002), 126.

13   Stewart Brand, "The Dawn of De-Extinction. Are You Ready?," (2013): http://longnow.org /revive/de-extinction/2013/stewart-brand-the-dawn-of-de-extinction-are-you-ready, accessed May 25, 2021; Elin Kelsey and Clayton Hanmer, *Not Your Typical Book about the Environment* (Toronto: Owlkids, 2010).

14   Head, *Hope and Grief*.

15   Ashlee Cunsolo and Neville R. Ellis, "Ecological Grief as a Mental Health Response to Climate Change–Related Loss," *Nature Climate Change* 8, no. 4 (2018).

16 https://theconversation.com/hope-and-mourning-in-the-anthropocene-understanding
    -ecological-grief-88630, accessed May 25, 2021.

17 Rose, *Wild Dog Dreaming*, 98.

18 James Hatley, "Blaspheming Humans: Levinasian Politics and the Cove," *Environmental Philosophy* 8, no. 2 (2011).

19 Deborah Bird Rose, "Slowly: Writing into the Anthropocene," *TEXT* 20 (2013).

20 Rose, *Wild Dog Dreaming*, 5.

21 My understanding here is influenced by Wendell Berry's important work that sees some political action as something moved less by the hope of "public success" than it is by "the hope of preserving qualities in one's own heart and spirit that would be destroyed by acquiescence." Wendell Berry, *What Are People For?: Essays* (Berkeley, CA: Counterpoint Press, 2010), 62. My own framing, however, emphasizes possibilities for relationship over the individual's cultivation of a particular self.

EPILOGUE

1 Joji Uchikawa et al., "Geochemical and Climate Modeling Evidence for Holocene Aridification in Hawaii: Dynamic Response to a Weakening Equatorial Cold Tongue," *Quaternary Science Reviews* 29, nos. 23–24 (2010).

2 Michael G. Hadfield, Stephen E. Miller, and Anne H. Carwile, "The Decimation of Endemic Hawaiian Tree Snails by Alien Predators," *American Zoologist* 33 (1993).

3 Jessica A. Garza et al., "Changes of the Prevailing Trade Winds over the Islands of Hawaii and the North Pacific," *Journal of Geophysical Research: Atmospheres* 117, no. D11 (2012): 16.

4 Charles H. Fletcher, *Hawai'i's Changing Climate: Briefing Sheet, 2010* (Honolulu: University of Hawai'i Sea Grant College Program, 2010), 2.

5 Fletcher, *Hawai'i's Changing Climate*, 3.

6 https://blogs.scientificamerican.com/extinction-countdown/snails-going-extinct, accessed May 25, 2021.

7 T. A. Norris et al., "An Integrated Approach for Assessing Translocation as an Effective Conservation Tool for Hawaiian Monk Seals," *Endangered Species Research* 32 (2017); Chris D. Thomas, "Translocation of Species, Climate Change, and the End of Trying to Recreate Past Ecological Communities," *Trends in Ecology and Evolution* 26 (2011).

8 Van Dooren, "Moving Birds in Hawai'i."

9 Yeung et al., "Overlooked but Not Forgotten."

10 Fujikane, *Mapping Abundance for a Planetary Future*; Goodyear-Ka'ōpua, Hussey, and Kahunawaika'ala, *A Nation Rising*. For related discussions in other parts of the world, see Davis

and Todd, "On the Importance of a Date, or Decolonizing the Anthropocene"; Whyte, "Our Ancestors' Dystopia Now"; DeLoughrey, *Allegories of the Anthropocene.*

11   Goodyear-Kaʻōpua, "Protectors of the Future," 186.

12   Kawika Winter et al., "The Moku System," 1.

13   Goodyear-Kaʻōpua, "Rebuilding the ʻAuwai"; Puaʻala Pascua et al., "Beyond Services: A Process and Framework to Incorporate Cultural, Genealogical, Place-Based, and Indigenous Relationships in Ecosystem Service Assessments," *Ecosystem Services* 26 (2017); Gon and Winter, "A Hawaiian Renaissance"; Vaughan, *Kaiaulu: Gathering Tides.*

Bibliography

Amundson, Ron. "John T. Gulick and the Active Organism: Adaptation, Isolation, and the Politics of Evolution." In *Darwin's Laboratory: Evolutionary Theory and Natural History in the Pacific*, edited by Roy MacLeod and Philip Rehbock, 110–139. Honolulu: University of Hawai'i Press, 1994.

Andelman, Sandy J., and William F. Fagan. "Umbrellas and Flagships: Efficient Conservation Surrogates or Expensive Mistakes?" *Proceedings of the National Academy of Sciences* 97 (2000): 5954–5959.

Anderson-Fung, Puanani O., and Kepā Maly. "Hawaiian Ecosystems and Culture: Why Growing Plants for Lei Helps to Preserve Hawai'i's Natural and Cultural Heritage." *CTAHR Resource Management Publication* 16 (2009).

Anonymous. "Mount Tantalus Got Its Name from Punahou Boys." *The Pacific Commercial Advertiser*, October 7, 1901, 3.

Anonymous, "Notes of the Week," *The Hawaiian Gazette*, October 8, 1873.

Anonymous. *One Last Effort to Save the Po'ouli*. Honolulu: Hawai'i Department of Land and Natural Resources, US Fish and Wildlife Service—Pacific Islands, and Zoological Society of San Diego, 2003.

Archer, Seth. *Sharks upon the Land: Colonialism, Indigenous Health, and Culture in Hawai'i, 1778–1855*. Cambridge: Cambridge University Press, 2018.

Arista, Noelani. "Listening to Leoiki: Engaging Sources in Hawaiian History." *Biography* 32, no. 1 (2009): 66–73.

Arista, Noelani. "Navigating Uncharted Oceans of Meaning: Kaona as Historical and Interpretive Method." *PMLA* 125, no. 3 (2010): 663–669.

Athens, J. Stephen. "*Rattus exulans* and the Catastrophic Disappearance of Hawai'i's Native Lowland Forest." *Biological Invasions* 11, no. 7 (2009): 1489–1501.

Baillie, J. E., C. Hilton-Taylor, and S. N. Stuart, eds. *A Global Species Assessment*. Gland, Switzerland: IUCN, 2004.

Baldwin, David D. "Descriptions of New Species of Achatinellidae from the Hawaiian Islands." *Proceedings of the Academy of Natural Sciences of Philadelphia* 47 (1895): 214–236.

Baldwin, David D. "Land Shell Collecting on Oahu." *Hawaii's Young People* 4, no. 8 (1900): 239–243.

Baldwin, David D. "The Land Shells of the Hawaiian Islands." In *Hawaiian Almanac and Annual*, 55–63. Honolulu: Press Publishing Company, 1887.

Baldy, Cutcha Risling. "Coyote Is Not a Metaphor: On Decolonizing, (Re)claiming and (Re) naming 'Coyote.'" *Decolonization: Indigeneity, Education & Society* 4, no. 1 (2015): 1–20.

Banivanua Mar, Tracey. *Decolonisation and the Pacific: Indigenous Globalisation and the Ends of Empire*. Cambridge: Cambridge University Press, 2016.

Banner, Stuart. *Possessing the Pacific: Land, Settlers, and Indigenous People from Australia to Alaska*. Cambridge, MA: Harvard University Press, 2007.

Barnett, Joshua Trey. "Naming, Mourning, and the Work of Earthly Coexistence." *Environmental Communication* 13, no. 3 (2019): 287–299.

Barnosky, Anthony D., Nicholas Matzke, Susumu Tomiya, Guinevere O. U. Wogan, Brian Swartz, Tiago B. Quental, Charles Marshall, Jenny L. McGuire, Emily L. Lindsey, Kaitlin C. Maguire, Ben Mersey, and Elizabeth A. Ferrer. "Has the Earth's Sixth Mass Extinction Already Arrived?" *Nature* 471 (2011): 51–57.

Bastian, Michelle. "Whale Falls, Suspended Ground, and Extinctions Never Known." *Environmental Humanities* 12, no. 2 (2020): 454–474.

Beckwith, Martha Warren (trans.) *The Kumulipo: A Hawaiian Creation Chant*. Chicago: University of Chicago Press, 1951.

Benson, Etienne S. "Naming the Ethological Subject." *Science in Context* 29, no. 1 (2016): 107–128.

Berger-Tal, Oded, Tal Polak, Aya Oron, Burt P. Kotler, and David Saltz. "Integrating Animal Behavior and Conservation Biology: A Conceptual Framework." *Behavioral Ecology* 22 (2011): 1–4.

Berry, Wendell. *What Are People For? Essays*. Berkeley, CA: Counterpoint Press, 2010.

Bezan, Sarah. "The Endling Taxidermy of Lonesome George: Iconographies of Extinction at the End of the Line." *Configurations* 27, no. 2 (2019): 211–238.

Blair-Stahn, Chai Kaiaka. "The Hula Dancer as Environmentalist: (Re-)Indigenising Sustainability through a Holistic Perspective on the Role of Nature in Hula." In *Proceedings of the 4th International Traditional Knowledge Conference*, edited by Joseph S. Te Rito, and Susan M. Healy, 60–64. Auckland: Knowledge Exchange Programme of Ngā Pae o te Māramatanga, 2010.

Borkfelt, Sune. "What's in a Name?: Consequences of Naming Non-Human Animals." *Animals* 1 (2011): 116–125.

Brand, Stewart. "The Dawn of De-Extinction. Are You Ready?" (2013): https://www.ted.com/talks/stewart_brand_the_dawn_of_de_extinction_are_you_ready?.

Brantlinger, Patrick. *Dark Vanishings: Discourse on the Extinction of Primitive Races, 1800–1930.* Ithaca, NY: Cornell University Press, 2003.

Brentari, Carlo. *Jakob von Uexküll: The Discovery of the Umwelt between Biosemiotics and Theoretical Biology.* Dordrecht: Springer, 2015.

Brossard, D., P. Belluck, F. Gould, and C. D. Wirz. "Promises and Perils of Gene Drives: Navigating the Communication of Complex, Post-Normal Science." *Proceedings of the National Academy of Sciences USA* 116, no. 16 (2019): 7692–7697.

Buchanan, Brett. "On the Trail of a Philosopher Ethologist." Paper presented at the *The History, Philosophy, and Future of Ethology IV.* Perth: Curtin University, 2019.

Buchanan, Brett, Michelle Bastian, and Matthew Chrulew. "Introduction: Field Philosophy and Other Experiments." *Parallax* 24, no. 4 (2018): 383–391.

Cameron, Robert. *Slugs and Snails.* London: HarperCollins UK, 2016.

Candland, Douglas K. "The Animal Mind and Conservation of Species: Knowing What Animals Know." *Current Science* 89, no. 7 (2005): 1122–1127.

Cardoso, Pedro, Terry L. Erwin, Paulo A. V. Borges, and Tim R. New. "The Seven Impediments in Invertebrate Conservation and How to Overcome Them." *Biological Conservation* 144, no. 11 (2011): 2647–2655.

Castree, Noel. "The Anthropocene and the Environmental Humanities: Extending the Conversation." *Environmental Humanities* 5 (2014): 233–260.

Chase, Ronald B. *Behavior and Its Neural Control in Gastropod Molluscs.* Oxford: Oxford University Press, 2002.

Christensen, Carl C. "Dr. Kondo Retires as Bishop Museum Malacologist." *Hawaiian Shell News* 29, no. 1 (1981): 1.

Christensen, Carl C. "Should We Open This Can of Worms? A Call for Caution Regarding Use of the Nematode *Phasmarhabditis hermaphrodita* for Control of Pest Slugs and Snails in the United States." *Tentacle* 27 (2019): 2–4.

Christensen, Carl C., and Michael G. Hadfield. *Field Survey of Endangered Oʻahu Tree Snails (Genus Achatinella) on the Makua Military Reservation.* Honolulu: Division of Malacology, Bernice P. Bishop Museum, 1984.

Chrulew, Matthew. "Reconstructing the Worlds of Wildlife: Uexküll, Hediger, and Beyond." *Biosemiotics* 13, no. 1 (2020): 137–149.

Clark, Nigel. "Geo-Politics and the Disaster of the Anthropocene." *The Sociological Review* 62 (2014): 19–37.

Clement, Russell. "From Cook to the 1840 Constitution: The Name Change from Sandwich to Hawaiian Islands." *Hawaiian Journal of History* 14 (1980): 50–57.

Clench, William J. "John T. Gulick's Hawaiian Land Shells." *Nautilus* 72 (1959): 95–98.

Clifford, Kavan T., Liaini Gross, Kwame Johnson, Khalil J. Martin, Nagma Shaheen, and Melissa A. Harrington. "Slime-Trail Tracking in the Predatory Snail, *Euglandina rosea.*" *Behavioral Neuroscience* 117, no. 5 (2003): 1086–1095.

Conniff, Richard. "Conservation Conundrum: Is Focusing on a Single Species a Good Strategy?" *Yale Environment 360* (2018).

Cook, A. "Homing by the Slug *Limax pseudoflavus.*" *Animal Behaviour* 27 (1979): 545–552.

Cooper, David E. *The Measure of Things: Humanism, Humility, and Mystery.* Oxford: Oxford University Press, 2007.

Costello, M. J., R. M. May, and N. E. Stork. "Can We Name Earth's Species before They Go Extinct?" *Science* 339, no. 6118 (2013): 413–416.

Cowie, Robert H. "Patterns of Introduction of Non-Indigenous Non-Marine Snails and Slugs in the Hawaiian Islands." *Biodiversity and Conservation* 7, no. 3 (1998): 349–368.

Cowie, Robert H. "Variation in Species Diversity and Shell Shape in Hawaiian Land Snails: In Situ Speciation and Ecological Relationships." *Evolution* 49, no. 6 (1995): 1191–1202.

Cowie, Robert H. "Yoshio Kondo: Bibliography and List of Taxa." *Bishop Museum Occasional Papers* 32 (1993).

Cowie, Robert H., Neal L. Evenhuis, and Carl C. Christensen. *Catalog of the Native Land and Freshwater Molluscs of the Hawaiian Islands.* Leiden, the Netherlands: Backhuys Publishers, 1995.

Cowie, Robert H., and Brenden S. Holland. "Molecular Biogeography and Diversification of the Endemic Terrestrial Fauna of the Hawaiian Islands." *Philosophical Transactions of the Royal Society of London B* 363, no. 1508 (2008): 3363–3376.

Cowie, Robert H., Claire Regnier, Benoit Fontaine, and Philippe Bouchet. "Measuring the Sixth Extinction: What Do Mollusks Tell Us?" *Nautilus* 1 (2017): 3–41.

Crist, Eileen. "Ecocide and the Extinction of Animal Minds." In *Ignoring Nature No More: The Case for Compassionate Conservation*, edited by Marc Bekoff, 45–61. Chicago: University of Chicago Press, 2013.

Cunsolo, Ashlee, and Neville R. Ellis. "Ecological Grief as a Mental Health Response to Climate Change-Related Loss." *Nature Climate Change* 8, no. 4 (2018): 275–281.

Dalsimer, Alison A. "Threatened and Endangered Species on DoD Lands." *Department of Defense Natural Resources Program Fact Sheet* (2016).

D'Arcy, Paul. *Transforming Hawai'i: Balancing Coercion and Consent in Eighteenth-Century Kānaka Maoli Statecraft.* Canberra: ANU Press, 2018.

Daston, Lorraine. "Type Specimens and Scientific Memory." *Critical Inquiry* 31 (2004): 153–182.

Davies, M. S., and J. Blackwell. "Energy Saving through Trail Following in a Marine Snail." *Proceedings of the Royal Society B* 274, no. 1614 (2007): 1233–1236.

Davis, C. J., and G. D. Butler. "Introduced Enemies of the Giant African Snail, *Achatina fulica* Bowdich, in Hawaii (Pulmonata: Achatinidae)." *Proceedings of the Hawaiian Entomological Society* 18, no. 3 (1964): 377–389.

Davis, Heather, and Zoe Todd. "On the Importance of a Date, or Decolonizing the Anthropocene." *ACME: An International E-Journal for Critical Geographies* 16, no. 4 (2017): 761–780.

Dawson, Ashley. *Extinction: A Radical History.* New York: OR Books, 2016.

DeLoughrey, Elizabeth M. *Allegories of the Anthropocene.* Durham, NC: Duke University Press, 2019.

DeLoughrey, Elizabeth M. "The Myth of Isolates: Ecosystem Ecologies in the Nuclear Pacific." *Cultural Geographies* 20, no. 2 (2013): 167–184.

Denny, M. "The Role of Gastropod Pedal Mucus in Locomotion." *Nature* 285 (1980): 160–161.

Despret, Vinciane. "The Enigma of the Raven." *Angelaki* 20 (2015): 57–72.

Despret, Vinciane. "It Is an Entire World That Has Disappeared." In *Extinction Studies: Stories of Time, Death and Generations,* edited by Deborah Bird Rose, Thom van Dooren, and Matthew Chrulew, 216–222. New York: Columbia University Press, 2017.

Donaldson, Michael R., Nicholas J. Burnett, Douglas C. Braun, Cory D. Suski, Scott G. Hinch, Steven J. Cooke, and Jeremy T. Kerr. "Taxonomic Bias and International Biodiversity Conservation Research." *FACETS* 1 (2016): 105–113.

Dubois, Alain. "Describing New Species." *Taprobanica: The Journal of Asian Biodiversity* 2, no. 1 (2011): 6–24.

Dubois, Alain. "The Relationships between Taxonomy and Conservation Biology in the Century of Extinctions." *C.R. Biologies* 326 Supplement 1 (2003): S9–S21.

Dundee, Dee S., Paul H. Phillips, and John D. Newsom. "Snails on Migratory Birds." *Nautilus* 80 (1967): 89–91.

Edelstam, Carl, and Carina Palmer. "Homing Behaviour in Gastropodes." *Oikos* 2 (1950): 259–270.

Eisenhauer, Nico, Aletta Bonn, and Carlos A. Guerra. "Recognizing the Quiet Extinction of Invertebrates." *Nature Communications* 10, no. 50 (2019): 1–3.

Evans, Kate. "Change Species Names to Honor Indigenous Peoples, Not Colonizers, Researchers Say." *Scientific American,* November 3, 2020. https://www.scientificamerican.com/article/change-species-names-to-honor-indigenous-peoples-not-colonizers-researchers-say.

Fermantez, Kali. "Re-Placing Hawaiians in dis Place We Call Home." *Hūlili: Multidisciplinary Research on Hawaiian Well-Being* 8 (2012): 97–131.

Findlen, Paula. *Possessing Nature: Museums, Collecting, and Scientific Culture in Early Modern Italy.* Los Angeles: University of California Press, 1994.

Fischer, J., and D. B. Lindenmayer. "An Assessment of the Published Results of Animal Relocations." *Biological Conservation* 96 (2000): 1–11.

Fischer, John Ryan. "Cattle in Hawai'i: Biological and Cultural Exchange." *Pacific Historical Review* 76, no. 3 (2007): 347–372.

Fletcher, Charles H. *Hawai'i's Changing Climate: Briefing Sheet, 2010*. Honolulu: University of Hawai'i Sea Grant College Program, 2010.

Fontaine, Benoît, Adrien Perrard, and Philippe Bouchet. "21 Years of Shelf Life between Discovery and Description of New Species." *Current Biology* 22 (2012): R943–R944.

Friese, Carrie. *Cloning Wild Life: Zoos, Captivity, and the Future of Endangered Animals*. New York: NYU Press, 2013.

Fujikane, Candace. *Mapping Abundance for a Planetary Future: Kanaka Maoli and Critical Settler Cartographies in Hawai'i*. Durham, NC: Duke University Press, 2021.

Gagné, Wayne C. "Conservation Priorities in Hawaiian Natural Systems." *BioScience* 38, no. 4 (1988): 264–271.

Gagné, Wayne C., and Carl C. Christensen. "Conservation Status of Native Terrestrial Invertebrates in Hawai'i." In *Hawaii's Terrestrial Ecosystems: Preservation and Management*, edited by C. P. Stone and J. M. Scott, 105–126. Honolulu: Cooperative National Park Resources Studies Unit, University of Hawai'i, 1985.

Galka, Jonathan. "*Liguus* Landscapes: Professional Malacology, Amateur Ligging, and the Social Life of Snail Science." *Journal of the History of Biology* 55, no. 4 (2022).

Galka, Jonathan. "Mollusk Loves: Becoming with Native and Introduced Land Snails in the Hawaiian Islands." *Island Studies Journal* (forthcoming).

Galla, Candace Kaleimamoowahinekapu, Louise Janet Leiola Aquino Galla, Dennis Kana'e Keawe, and Larry Lindsey Kimura. "Perpetuating Hula: Globalization and the Traditional Art." *Pacific Arts Association* 14, no. 1 (2015): 129–140.

Garza, Jessica A., Pao-Shin Chu, Chase W. Norton, and Thomas A. Schroeder. "Changes of the Prevailing Trade Winds over the Islands of Hawaii and the North Pacific." *Journal of Geophysical Research: Atmospheres* 117, no. D11 (2012): 1–18.

Ghosh, Amitav. *The Nutmeg's Curse: Parables for a Planet in Crisis*. Chicago: University of Chicago Press, 2021.

Giggs, Rebecca. *Fathoms: The World in the Whale*. London: Scribe, 2020.

Goldberg-Hiller, Jonathan, and Noenoe K. Silva. "Sharks and Pigs: Animating Hawaiian Sovereignty against the Anthropological Machine." *The South Atlantic Quarterly* 110 (2011): 429–446.

Gomes, Noah. "Reclaiming Native Hawaiian Knowledge Represented in Bird Taxonomies." *Ethnobiology Letters* 11, no. 2 (2020): 30–43.

Gon, Sam 'Ohu, and Kawika Winter. "A Hawaiian Renaissance That Could Save the World." *American Scientist* 107, no. 4 (2019): 232–239.

Gonzalez, Vernadette Vicuna. *Securing Paradise: Tourism and Militarism in Hawai'i and the Philippines*. Durham, NC: Duke University Press, 2013.

González-Oreja, José Antonio. "The Encyclopedia of Life vs. the Brochure of Life: Exploring the Relationships between the Extinction of Species and the Inventory of Life on Earth." *Zootaxa* 1965, no. 1 (2008): 61–68.

Goodyear-Kaʻōpua, Noelani. "Protectors of the Future, Not Protestors of the Past: Indigenous Pacific Activism and Mauna a Wākea." *South Atlantic Quarterly* 116, no. 1 (2017): 184–194.

Goodyear-Kaʻōpua, Noelani. "Rebuilding the ʻAuwai: Connecting Ecology, Economy and Education in Hawaiian Schools." *AlterNative: An International Journal of Indigenous Peoples* 5, no. 2 (2009): 46–77.

Goodyear-Kaʻōpua, Noelani, Ikaika Hussey, and Erin Kahunawaikaʻala, eds. *A Nation Rising: Hawaiian Movements for Life, Land, and Sovereignty* Durham, NC: Duke University Press, 2014.

Griffiths, Tom. *Hunters and Collectors: The Antiquarian Imagination in Australia.* Cambridge: Cambridge University Press, 1996.

Griffiths, Tom, and Libby Robin, eds. *Ecology and Empire: Environmental History of Settler Societies.* Seattle: University of Washington Press, 1997.

Gulick, Addison. *Evolutionist and Missionary, John Thomas Gulick: Portrayed through Documents and Discussions.* Chicago: University of Chicago Press, 1932.

Hadfield, Michael G. "Extinction in Hawaiian Achatinelline Snails." *Malacologia* 27, no. 1 (1986): 67–81.

Hadfield, Michael. "Snails That Kill Snails." In *The Feral Atlas,* edited by Anna L. Tsing, Jennifer Deger, Alder Keleman Saxena, and Feifei Zhou. Stanford, CA: Stanford University Press, 2021.

Hadfield, Michael G., and Donna Haraway. "The Tree-Snail Manifesto." *Current Anthropology* 60, no. S20 (2019): S209–S235.

Hadfield, Michael G., Stephen E. Miller, and Anne H. Carwile. "The Decimation of Endemic Hawaiian Tree Snails by Alien Predators." *American Zoologist* 33 (1993): 610–622.

Hadfield, Michael G., and Barbara Shank Mountain. "A Field Study of a Vanishing Species, *Achatinella mustelina* (Gastropoda, Pulmonata), in the Waianae Mountains of Oahu." *Pacific Science* 34, no. 4 (1980): 345–358.

Haraway, Donna. "Anthropocene, Capitalocene, Plantationocene, Chthulucene: Making Kin." *Environmental Humanities* 6 (2015): 159–165.

Haraway, Donna. *When Species Meet.* Minneapolis: University of Minnesota Press, 2008.

Hart, Alan D. "Living Jewels Imperiled." *Defenders* 50 (1975): 482–486.

Hart, Alan D. "The Onslaught against Hawaii's Tree Snails." *Natural History* 87 (1978): 46–57.

Hatley, James. "Blaspheming Humans: Levinasian Politics and the Cove." *Environmental Philosophy* 8, no. 2 (2011): 1–21.

Hauʻofa, Epeli. "Our Sea of Islands." In *A New Oceania: Rediscovering Our Sea of Islands,* edited by Eric Waddell, Vijay Naidu, and Epeli Hauʻofa, 2–16. Suva: University of the South Pacific, 1993.

Havlick, David. *Bombs Away: Militarization, Conservation, and Ecological Restoration*. Chicago: University of Chicago Press, 2018.

Havlick, David. "Logics of Change for Military-to-Wildlife Conversions in the United States." *GeoJournal* 69, no. 3 (2007): 151–164.

Hayes, Kenneth A., John Slapcinsky, David R. Sischo, Jaynee R. Kim, and Norine W. Yeung. "The Last Known *Endodonta* Species? *Endodonta christenseni* sp. nov. (Gastropoda: Endodontidae)." *Bishop Museum Occasional Papers* 138 (2020): 1–15.

Hayward, Eva. "Sensational Jellyfish: Aquarium Affects and the Matter of Immersion." *differences* 23, no. 3 (2012): 161–196.

Head, Lesley. "The Anthropoceneans." *Geographical Research* 53, no. 3 (2015): 313–320.

Head, Lesley. *Hope and Grief in the Anthropocene: Re-Conceptualising Human–Nature Relations*. London: Routledge, 2016.Heaney, Lawrence R. "Is a New Paradigm Emerging for Oceanic Island Biogeography?," *Journal of Biogeography* 34 (2007): 753–757.

Heard, Stephen B. *Charles Darwin's Barnacle and David Bowie's Spider: How Scientific Names Celebrate Adventurers, Heroes, and Even a Few Scoundrels*. New Haven, CT: Yale University Press, 2020.

Heise, Ursula K. *Imagining Extinction: The Cultural Meanings of Endangered Species*. Chicago: University of Chicago Press, 2016.

Holland, Brenden S. "Land Snails." In *Encyclopedia of Islands*, 537–542. Los Angeles: University of California Press, 2009.

Holland, Brenden S., Luciano M. Chiaverano, and Cierra K. Howard. "Diminished Fitness in an Endemic Hawaiian Snail in Nonnative Host Plants." *Ethology Ecology & Evolution* 29, no. 3 (2017): 229–240.

Holland, Brenden S., Taylor Chock, Alan Lee, and Shinji Sugiura. "Tracking Behavior in the Snail *Euglandina rosea*: First Evidence of Preference for Endemic vs. Biocontrol Target Pest Species in Hawai'i." *American Malacological Bulletin* 30, no. 1 (2012): 153–157.

Holland, Brenden S., and Robert H. Cowie. "A Geographic Mosaic of Passive Dispersal: Population Structure in the Endemic Hawaiian Amber Snail *Succinea caduca* (Mighels, 1845)." *Molecular Ecology* 16, no. 12 (2007): 2422–2435.

Holland, Brenden S., Marianne Gousy-Leblanc, and Joanne Y. Yew. "Strangers in the Dark: Behavioral and Biochemical Evidence for Trail Pheromones in Hawaiian Tree Snails." *Invertebrate Biology* 137, no. 2 (2018): 124–132.

Holland, Brenden S., and Michael G. Hadfield. "Islands within an Island: Phylogeography and Conservation Genetics of the Endangered Hawaiian Tree Snail *Achatinella mustelina*." *Molecular Ecology* 11, no. 3 (2002): 365–375.

ho'omanawanui, ku'ualoha. *Voices of Fire: Reweaving the Literary Lei of Pele and Hi'iaka*. Minneapolis: University of Minnesota Press, 2014.

Howes, David, ed. *Empire of the Senses: The Sensual Culture Reader* Oxford: Berg Publishers, 2004.

Hunt, Terry L. "Rethinking Easter Island's Ecological Catastrophe." *Journal of Archaeological Science* 34, no. 3 (2007): 485–502.

Ingold, Tim. "An Anthropologist Looks at Biology." *Man (N.S.)* 25 (1990): 208–229.

James, A. F., R. Brown, K. A. Weston, and K. Walker. "Modelling the Growth and Population Dynamics of the Exiled Stockton Coal Plateau Landsnail, *Powelliphanta augusta*." *New Zealand Journal of Zoology* 40, no. 3 (2013): 175–185.

Jardine, Nicholas, James A. Secord, and Emma C. Spary, eds. *Cultures of Natural History*. Cambridge: Cambridge University Press, 1996.

Johnson, Elizabeth, Harlan Morehouse, Simon Dalby, Jessi Lehman, Sara Nelson, Rory Rowan, Stephanie Wakefield, and Kathryn Yusoff. "After the Anthropocene: Politics and Geographic Inquiry for a New Epoch." *Progress in Human Geography* 38, no. 3 (2014): 439–456.

Jørgensen, Dolly. "Endling: The Power of the Last in an Extinction-Prone World." *Environmental Philosophy* 14, no. 1 (2017): 119–138.

Jørgensen, Dolly. *Recovering Lost Species in the Modern Age: Histories of Longing and Belonging*. Cambridge, MA: MIT Press, 2019.

Kauanui, J. Kēhaulani. *Hawaiian Blood: Colonialism and the Politics of Sovereignty and Indigeneity*. Durham, NC: Duke University Press, 2008.

Kauanui, J. Kēhaulani. *Paradoxes of Hawaiian Sovereignty: Land, Sex, and the Colonial Politics of State Nationalism*. Durham, NC: Duke University Press, 2018.

Kay, E. Alison. "Missionary Contributions to Hawaiian Natural History: What Darwin Didn't Know." *The Hawaiian Journal of History* 31 (1997): 27–52.

Kelly, Marion, and Nancy Aleck. *Mākua Means Parents: A Brief Cultural History of Mākua Valley*. Honolulu: American Friends Service Committee, 1997.

Kelsey, Elin, and Clayton Hanmer. *Not Your Typical Book about the Environment*. Toronto: Owlkids, 2010.

Kimmerer, Robin Wall. *Braiding Sweetgrass: Indigenous Wisdom, Scientific Knowledge and the Teachings of Plants*. Minneapolis, MN: Milkweed Editions, 2013.

Kirch, Patrick V. "When Did the Polynesians Settle Hawai'i? A Review of 150 Years of Scholarly Inquiry and a Tentative Answer." *Hawaiian Archaeology* 12 (2011): 3–26.

Kirksey, S. Eben. *Emergent Ecologies*. Durham, NC: Duke University Press, 2015.

Kobayashi, Sharon R., and Michael G. Hadfield. "An Experimental Study of Growth and Reproduction in the Hawaiian Tree Snails *Achatinella mustelina* and *Partulina redfieldii* (Achatinellinae)." *Pacific Science* 50, no. 4 (1996): 339–354.

Koene, Joris M., and Andries Ter Maat. "Coolidge Effect in Pond Snails: Male Motivation in a Simultaneous Hermaphrodite." *BMC Evolutionary Biology* 7, no. 1 (2007): 1–6.

Kohler, Robert E. *All Creatures: Naturalists, Collectors, and Biodiversity, 1850–1950*. Princeton, NJ: Princeton University Press, 2013.

Kohler, Robert E. "Finders, Keepers: Collecting Sciences and Collecting Practice." *History of Science* 45, no. 4 (2007): 428–454.

Kondo, Yoshio, and William J. Clench. "Charles Montague Cooke, Jr.: A Bio-Bibliography." *Bernice P. Bishop Museum, Special Publication* 42 (1952): 1–56.

Leonard, David L., Jr. "Recovery Expenditures for Birds Listed under the US Endangered Species Act: The Disparity between Mainland and Hawaiian Taxa." *Biological Conservation* 141 (2008): 2054–2061.

Leopold, Aldo. *A Sand County Almanac, and Sketches Here and There*. New York: Oxford University Press, 1949.

Lewontin, Richard. *The Triple Helix: Gene, Organism, and Environment*. Cambridge, MA: Harvard University Press, 2000.

Lorimer, Jamie. *Wildlife in the Anthropocene: Conservation after Nature*. Minneapolis: University of Minnesota Press, 2015.

Lukowiak, K., H. Sunada, M. Teskey, K. Lukowiak, and S. Dalesman. "Environmentally Relevant Stressors Alter Memory Formation in the Pond Snail *Lymnaea*." *Journal of Experimental Biology* 217 (2014): 76–83.

Lydeard, Charles, Robert H. Cowie, Winston F. Ponder, Arthur E. Bogan, Philippe Bouchet, Stephanie A. Clark, Kevin S. Cummings, Terrence J. Frest, Olivier Gargominy, and Dai G. Herbert. "The Global Decline of Nonmarine Mollusks." *BioScience* 54, no. 4 (2004): 321–330.

Macfarlane, Robert. *Underland: A Deep Time Journey*. London: Penguin, 2019.

MacLennan, Carol A. *Sovereign Sugar: Industry and Environment in Hawai'i*. Honolulu: University of Hawai'i Press, 2014.

MacLeod, Roy M., and Philip F. Rehbock, eds. *Darwin's Laboratory: Evolutionary Theory and Natural History in the Pacific*. Honolulu: University of Hawai'i Press, 1994.

Makua Implementation Team. *Implementation Plan: Makua Military, Reservation, Island of Oahu*. Honolulu: United States Army Garrison, Hawai'i, 2003.

Mawyer, Alexander, and Jerry K. Jacka. "Sovereignty, Conservation and Island Ecological Futures." *Environmental Conservation* 45 no. 3 (2018): 238–251.

McCarthy, Niall. "Report: The U.S. Military Emits More CO2 than Many Industrialized Nations." *Forbes*, June 13, 2019.

McDougall, Brandy Nālani. *Finding Meaning: Kaona and Contemporary Hawaiian Literature*. Tucson: University of Arizona Press, 2016.

McDougall, Brandy Nālani, and Georganne Nordstrom. "Ma Ka Hana Ka 'Ike (In the Work is the Knowledge): Kaona as Rhetorical Action." *College Composition and Communication* 63, no. 1 (2011): 98–121.

McGregor, Davianna Pomaika'i. "Waipi'o Valley, a Cultural *Kīpuka* in Early 20th Century Hawai'i." *The Journal of Pacific History* 30, no. 2 (1995): 194–209.

McHugh, Susan. *Love in a Time of Slaughters: Human–Animal Stories against Genocides and Extinctions*. University Park: Penn State University Press, 2019.

Mertl, Melissa. "Taxonomy in Danger of Extinction." *Science Magazine*, May 22, 2002.

Meyer, Wallace M., and Robert H. Cowie. "Feeding Preferences of Two Predatory Snails Introduced to Hawai'i and Their Conservation Implications." *Malacologia* 53, no. 1 (2010): 135–144.

Meyer, Wallace M., Rebecca Ostertag, and Robert H. Cowie. "Influence of Terrestrial Molluscs on Litter Decomposition and Nutrient Release in a Hawaiian Rain Forest." *Biotropica* 45, no. 6 (2013): 719–727.

Meyer, Wallace M., Norine W. Yeung, John Slapcinsky, and Kenneth A. Hayes. "Two for One: Inadvertent Introduction of *Euglandina* Species during Failed Bio-Control Efforts in Hawai'i." *Biological Invasions* 19, no. 5 (2017): 1399–1405.

Mitchell, Audra. "Revitalizing Laws, (Re)-Making Treaties, Dismantling Violence: Indigenous Resurgence against 'the Sixth Mass Extinction.'" *Social & Cultural Geography* 21, no. 7 (2020): 909–924.

Mora, C., A. Rollo, and D. P. Tittensor. "Comment on 'Can We Name Earth's Species before They Go Extinct?'" *Science* 341, no. 6143 (2013): 237.

Mora, C., D. P. Tittensor, S. Adl, A. G. Simpson, and B. Worm. "How Many Species Are There on Earth and in the Ocean?" *PLoS Biology* 9, no. 8 (2011): e1001127.

Mouffe, Chantal, and Ernesto Laclau. "Hope, Passion, Politics." In *Hope: New Philosophies for Change*, edited by Mary Zournazi, 122–148. Annandale, NSW: Pluto Press, 2002.

Nassar, Dalia, and Margaret M. Barbour. "Rooted." *Aeon*, October 16, 2019.

Naughton, Eileen Momilani. "The Bernice Pauahi Bishop Museum: A Case Study Analysis of Mana as a Form of Spiritual Communication in the Museum Setting." PhD dissertation, Simon Fraser University, 2001.

New, Timothy R. "Angels on a Pin: Dimensions of the Crisis in Invertebrate Conservation." *American Zoologist* 33, no. 6 (1993): 623–630.

Newell, Jennifer. *Trading Nature: Tahitians, Europeans and Ecological Exchange*. Honolulu: University of Hawai'i Press, 2010.

Ng, T. P., S. H. Saltin, M. S. Davies, K. Johannesson, R. Stafford, and G. A. Williams. "Snails and Their Trails: The Multiple Functions of Trail-Following in Gastropods." *Biological Reviews* 88, no. 3 (2013): 683–700.

Niheu, Kalamaoka'aina. "Pu'uhonua: Sanctuary and Struggle at Mākua." In *A Nation Rising: Hawaiian Movements for Life, Land, and Sovereignty*, edited by Noelani Goodyear-Ka'ōpua, Ikaika Hussey, and Erin Kahunawaika'ala, 161–179. Durham, NC: Duke University Press, 2014.

Nijhuis, Michelle. *Beloved Beasts: Fighting for Life in an Age of Extinction*. New York: W. W. Norton, 2021.

Nixon, Rob. *Slow Violence and the Environmentalism of the Poor*. Cambridge, MA: Harvard University Press, 2011.

Nogelmeier, Puakea. "Mai Pa'a I Ka Leo: Historical Voice in Hawaiian Primary Materials, Looking Forward and Listening Back." PhD dissertation, University of Hawai'i at Mānoa, 2003.

Norris, T. A., C. L. Littnan, F. M. D. Gulland, J. D. Baker, and J. T. Harvey. "An Integrated Approach for Assessing Translocation as an Effective Conservation Tool for Hawaiian Monk Seals." *Endangered Species Research* 32 (2017): 103–115.

Olson, Storrs L., and Helen F. James. "Descriptions of Thirty-Two New Species of Birds from the Hawaiian Islands: Part I. Non-Passeriformes." *Ornithological Monographs* 45 (1991): 1–88.

O'Rorke, R., G. M. Cobian, B. S. Holland, M. R. Price, V. Costello, and A. S. Amend. "Dining Local: The Microbial Diet of a Snail That Grazes Microbial Communities Is Geographically Structured." *Environmental Microbiology* 17, no. 5 (2015): 1753–1764.

O'Rorke, R., L. Tooman, K. Gaughen, B. S. Holland, and A. S. Amend. "Not Just Browsing: An Animal That Grazes Phyllosphere Microbes Facilitates Community Heterogeneity." *ISME Journal* 11, no. 8 (2017): 1788–1798.

Osorio, Jonathan Kamakawiwo'ole. "Hawaiian Souls: The Movement to Stop the U.S. Military Bombing of Kaho'olawe." In *A Nation Rising: Hawaiian Movements for Life, Land, and Sovereignty*, edited by Noelani Goodyear-Ka'ōpua, Ikaika Hussey, and Erin Kahunawaika'ala, 137–160. Durham, NC: Duke University Press, 2014.

Osorio, Jonathan Kay Kamakawiwo'ole. *Dismembering Lāhui: A History of the Hawaiian Nation to 1887*. Honolulu: University of Hawai'i Press, 2002.

Østmo, Liv, and John Law. "Mis/translation, Colonialism, and Environmental Conflict." *Environmental Humanities* 10, no. 2 (2018): 349–369.

Oyama, Susan. *Evolution's Eye: A Systems View of the Biology-Culture Divide*. Durham, NC: Duke University Press, 2000.

Ożgo, Małgorzata, Aydin Örstan, Małgorzata Kirschenstein, and Robert Cameron. "Dispersal of Land Snails by Sea Storms." *Journal of Molluscan Studies* 82, no. 2 (2016): 341–343.

Pascua, Pua'ala, Heather McMillen, Tamara Ticktin, Mehana Vaughan, and Kawika B. Winter. "Beyond Services: A Process and Framework to Incorporate Cultural, Genealogical, Place-Based, and Indigenous Relationships in Ecosystem Service Assessments." *Ecosystem Services* 26 (2017): 465–475.

Peralto, Leon No'eau. "O Koholālele, He 'Āina, He Kanaka, He I'a Nui Nona Ka Lā: Re-Membering Knowledge of Place in Koholālele, Hāmākua, Hawai'i." In *I Ulu I Ka 'Āina: Land*, edited by Jonathan Osorio, 76–98. Honolulu: University of Hawai'i Press, 2014.

Pisbry, Henry A., and C. Montague Cooke. *Manual of Conchology*, vol. 22. Philadelphia: Academy of Natural Sciences, 1912.

Plater, Zygmunt J. B. "In the Wake of the Snail Darter: An Environmental Law Paradigm and Its Consequences." *Journal of Law Reform* 19, no. 4 (1986): 805–862.

Plumwood, Val. *Environmental Culture: The Ecological Crisis of Reason*. London: Routledge, 2002.

Plumwood, Val. "Nature in the Active Voice." *Australian Humanities Review* 46 (2009): 113–129.

Powell, Miles A. *Vanishing America: Species Extinction, Racial Peril, and the Origins of Conservation.* Cambridge, MA: Harvard University Press, 2016.

Prather, C. M., S. L. Pelini, A. Laws, E. Rivest, M. Woltz, C. P. Bloch, I. Del Toro, C. K. Ho, J. Kominoski, T. A. Newbold, S. Parsons, and A. Joern. "Invertebrates, Ecosystem Services and Climate Change." *Biological Reviews* 88, no. 2 (2013): 327–348.

Primack, Richard. *Essentials of Conservation Biology.* Sunderland, MA: Sinauer Associates, 1993.

Purvis, Andy. "A Million Threatened Species? Thirteen Questions and Answers." (2019): https://ipbes.net/news/million-threatened-species-thirteen-questions-answers.

Quammen, David. *The Song of the Dodo: Island Biogeography in an Age of Extinctions.* New York: Scribner, 1996.

Radin, Joanna, and Emma Kowal, eds. *Cryopolitics: Frozen Life in a Melting World.* Cambridge, MA: MIT Press, 2017.

Reardon, Jenny. *Race to the Finish: Identity and Governance in an Age of Genomics.* Princeton, NJ: Princeton University Press, 2005.

Rees, W. J. "The Aerial Dispersal of Mollusca." *Journal of Molluscan Studies* 36, no. 5 (1965): 269–282.

Régnier, C., P. Bouchet, K. A. Hayes, N. W. Yeung, C. C. Christensen, D. J. Chung, B. Fontaine, and R. H. Cowie. "Extinction in a Hyperdiverse Endemic Hawaiian Land Snail Family and Implications for the Underestimation of Invertebrate Extinction." *Conservation Biology* 29, no. 6 (2015): 1715–1723.

Régnier, Claire, Benoît Fontaine, and Philippe Bouchet. "Not Knowing, Not Recording, Not Listing: Numerous Unnoticed Mollusk Extinctions." *Conservation Biology* 23, no. 5 (2009): 1214–1221.

Régnier, C., O. Gargominy, G. Falkner, and N. Puillandre. "Foot Mucus Stored on FTA® Cards Is a Reliable and Non-Invasive Source of DNA for Genetics Studies in Molluscs." *Conservation Genetics Resources* 3, no. 2 (2011): 377–382.

Richling, Ira, and Philippe Bouchet. "Extinct Even before Scientific Recognition: A Remarkable Radiation of Helicinid Snails (Helicinidae) on the Gambier Islands, French Polynesia." *Biodiversity and Conservation* 22, no. 11 (2013): 2433–2468.

Robin, Libby. "Ecology: A Science of Empire." In *Ecology and Empire: Environmental History of Settler Societies,* edited by Tom Griffiths, and Libby Robin, 63–75. South Carlton: Melbourne University Press, 1997.

Rohrer, Jobriath, Vincent Costello, Jamie Tanino, Lalasia Bialic-Murphy, Michelle Akamine, Jonathan Sprague, Stephanie Joe, and Clifford Smith. *Development of Tree Snail Protection Enclosures: From Design to Implementation.* Honolulu: Pacific Cooperative Studies Unit, University of Hawaiʻi at Mānoa, 2016.

Rooke, T. C. B. "The Sandwich Island Institute: Inaugural Thesis, Delivered before the Sandwich Island Institute, Dec. 12, 1838." *The Hawaiian Spectator*, 1838.

Rose, Deborah Bird. "Pattern, Connection, Desire: In Honour of Gregory Bateson." *Australian Humanities Review* 35 (2005).

Rose, Deborah Bird. *Reports from a Wild Country: Ethics for Decolonisation*. Sydney: UNSW Press, 2004.

Rose, Deborah Bird. "Slowly ~ Writing into the Anthropocene." *TEXT* 20 (2013): 1–14.

Rose, Deborah Bird. *Wild Dog Dreaming: Love and Extinction*. Charlottesville: University of Virginia Press, 2011.

Rose, Deborah Bird, and Thom van Dooren. "Unloved Others: Death of the Disregarded in the Time of Extinctions." *Australian Humanities Review* 50 (2011).

Rose, Deborah Bird, Thom van Dooren, and Matthew Chrulew. *Extinction Studies: Stories of Time, Death and Generations*. New York: Columbia University Press, 2017.

Rubis, June Mary, and Noah Theriault. "Concealing Protocols: Conservation, Indigenous Survivance, and the Dilemmas of Visibility." *Social & Cultural Geography* 21, no. 7 (2019): 1–23.

Rundell, Rebecca J. "Snails on an Evolutionary Tree: Gulick, Speciation, and Isolation." *American Malacological Bulletin* 29, nos. 1–2 (2011): 145–157.

Sagan, Dorian. "Samuel Butler's Willful Machines." *The Common Review* 9, no. 1 (2011): 10–19.

Sánchez-Bayo, Francisco, and Kris A. G. Wyckhuys. "Worldwide Decline of the Entomofauna: A Review of Its Drivers." *Biological Conservation* 232 (2019): 8–27.

Sartori, A. F., O. Gargominy, and B. Fontaine. "Anthropogenic Extinction of Pacific Land Snails: A Case Study of Rurutu, French Polynesia, with Description of Eight New Species of Endodontids (Pulmonata)." *Zootaxa* 3640 (2013): 343–372.

Sato, Aimee You, Melissa Renae Price, and Mehana Blaich Vaughan. "Kāhuli: Uncovering Indigenous Ecological Knowledge to Conserve Endangered Hawaiian Land Snails." *Society & Natural Resources* 31, no. 3 (2018): 320–334.

Sepkoski, David. *Catastrophic Thinking: Extinction and the Value of Diversity from Darwin to the Anthropocene*. Chicago: University of Chicago Press, 2020.

Severns, Mike. *Shells of the Hawaiian Islands: The Land Shells*. Hackenheim: ConchBooks, 2011.

Shaheen, Nagma, Kinjal Patel, Priyanka Patel, Michael Moore, and Melissa A. Harrington. "A Predatory Snail Distinguishes between Conspecific and Heterospecific Snails and Trails Based on Chemical Cues in Slime." *Animal Behaviour* 70, no. 5 (2005): 1067–1077.

Shehadeh, Raja. *Palestinian Walks: Notes on a Vanishing Landscape*. London: Profile Books, 2008.

Silva, Noenoe K. *Aloha Betrayed: Native Hawaiian Resistance to American Colonialism*. Durham, NC: Duke University Press, 2004.

Silva, Noenoe K. *The Power of the Steel-Tipped Pen: Reconstructing Native Hawaiian Intellectual History*. Durham, NC: Duke University Press, 2017.

Simpson, Inga. "Encounters with Amnesia: Confronting the Ghosts of Australian Landscape." *Griffith Review* 63 (2019).

Smith, Linda Tuhiwai. *Decolonizing Methodologies: Research and Indigenous Peoples*. London: Zed Books, 2012.

Snyder, Gary. *The Practice of the Wild*. Berkeley, CA: Counterpoint Press, 2010.

Solem, Alan. "How Many Hawaiian Land Snail Species Are Left? And What We Can Do for Them?" *Bishop Museum Occasional Papers* 30 (1990): 27–40.

Solnit, Rebecca. *Hope in the Dark: Untold Histories, Wild Possibilities*. Edinburgh: Canongate Books, 2016.

Stein, Bruce A., Cameron Scott, and Nancy Benton. "Federal Lands and Endangered Species: The Role of Military and Other Federal Lands in Sustaining Biodiversity." *BioScience* 58, no. 4 (2008): 339–347.

Stewart, Charles Samuel. "Addenda: *Achatina stewartii* and *A. oahuensis* [1823]." *Manual of Conchology* 22 (1912): 404–407.

Stillman, Amy Kuʻuleialoha. "The Hawaiian *Hula* and Legacies of Institutionalization." *Comparative American Studies: An International Journal* 5, no. 2 (2007): 221–234.

Stotz, Karola. "Extended Evolutionary Psychology: The Importance of Transgenerational Developmental Plasticity." *Frontiers in Psychology* 5, no. 908 (2014): 1–4.

Stuart, S. N., E. O. Wilson, J. A. McNeely, R. A. Mittermeier, and J. P. Rodríguez. "The Barometer of Life." *Science* 328, no. 5975 (2010): 177.

Takacs, David. *The Idea of Biodiversity: Philosophies of Paradise*. Baltimore: Johns Hopkins University Press, 1996.

Tengan, Ty P. Kāwika. *Native Men Remade: Gender and Nation in Contemporary Hawaiʻi*. Durham, NC: Duke University Press, 2008.

Territory of Hawaiʻi. *Report of the Commissioner of Agriculture and Forestry 1902*. Honolulu: Gazette Press, 1903.

Thomas, Chris D. "Translocation of Species, Climate Change, and the End of Trying to Recreate Past Ecological Communities." *Trends in Ecology and Evolution* 26 (2011): 216–221.

Trask, Haunani-Kay. "The Birth of the Modern Hawaiian Movement: Kalama Valley, Oʻahu." *Hawaiian Journal of History* 21 (1987): 126–153.

Trask, Haunani-Kay. *From a Native Daughter: Colonialism and Sovereignty in Hawaiʻi*. Honolulu: University of Hawaii Press, 1999.

Tsing, Anna Lowenhaupt. "Blasted Landscapes, and the Gentle Art of Mushroom Picking." In *The Multispecies Salon*, edited by Eben Kirksey. Durham, NC: Duke University Press, 2016.

Tsing, Anna Lowenhaupt. *Friction: An Ethnography of Global Connection*. Princeton, NJ: Princeton University Press, 2005.

Tummons, Patricia. "Army Tenure at Makua Valley Solidified after Statehood." *Environment Hawai'i* 3, no. 5 (1992).

Tummons, Patricia. "Army's Application for EPA Permit Is Long, but Not Informative." *Environment Hawai'i* 3, no. 5 (1992).

Tummons, Patricia. "Endangered Snails of Makua Valley Are Placed at Risk by Army Fires." *Environment Hawai'i* 3, no. 5 (1992).

Tummons, Patricia. "First the Cattle, then the Bombs, Oust Hawaiians from Makua Valley." *Environment Hawai'i* 3, no. 5 (1992).

Tummons, Patricia. "Terrestrial Ecosystems." In *The Value of Hawai'i: Knowing the Past, Shaping the Future*, edited by Craig Howes and Jonathan K. K. Osorio. Honolulu: University of Hawai'i Press, 2010.

Uchikawa, Joji, Brian N. Popp, Jane E. Schoonmaker, Axel Timmermann, and Stephan J. Lorenz. "Geochemical and Climate Modeling Evidence for Holocene Aridification in Hawai'i: Dynamic Response to a Weakening Equatorial Cold Tongue." *Quaternary Science Reviews* 29, nos. 23–24 (2010): 3057–3066.

US Department of Defense and US Fish & Wildlife Service. "The Military and the Endangered Species Act: Interagency Cooperation." *Factsheet* (2001).

Vagvolgyi, Joseph. "Body Size, Aerial Dispersal, and Origin of the Pacific Land Snail Fauna." *Systematic Biology* 24, no. 4 (1975): 465–488.

van Dooren, Thom. "Invasive Species in Penguin Worlds: An Ethical Taxonomy of Killing for Conservation." *Conservation and Society* 9 (2011): 286–298.

van Dooren, Thom. *Flight Ways: Life and Loss at the Edge of Extinction*. New York: Columbia University Press, 2014.

van Dooren, Thom. "Moving Birds in Hawai'i: Assisted Colonisation in a Colonised Land." *Cultural Studies Review* 25, no. 1 (2019): 41–64.

van Dooren, Thom. *The Wake of Crows: Living and Dying in Shared Worlds*. New York: Columbia University Press, 2019.

van Dooren, Thom, and Deborah Bird Rose. "Lively Ethography: Storying Animist Worlds." *Environmental Humanities* 8 (2016): 1–17.

Vaughan, Mehana Blaich. *Kaiaulu: Gathering Tides*. Corvallis: Oregon State University Press, 2018.

Vincent, Eve, and Timothy Neale, eds. *Unstable Relations: Indigenous People and Environmentalism in Contemporary Australia*. Perth: UWA Publishing, 2016.

Vine, David. *Base Nation: How US Military Bases Abroad Harm America and the World*. New York: Henry Holt and Company, 2015.

Vizenor, Gerald. *Survivance: Narratives of Native Presence.* Lincoln: University of Nebraska Press, 2008.

von Uexküll, Jakob. *A Foray into the Worlds of Animals and Humans: With a Theory of Meaning.* Translated by Joseph D. O'Neil. Minneapolis: University of Minnesota Press, 2010.

Wada, Shinichiro, Kazuto Kawakami, and Satoshi Chiba. "Snails Can Survive Passage through a Bird's Digestive System." *Journal of Biogeography* 39, no. 1 (2012): 69–73.

Welch, d'Alté A. "Distribution and Variation of *Achatinella mustelina* Mighels in the Waianae Mountains, Oahu." *Bernice P. Bishop Museum Bulletin* 152 (1938): 1–164.

West, Paige, James Igoe, and Dan Brockington. "Parks and Peoples: The Social Impact of Protected Areas." *Annual Review of Anthropology* 35 (2006): 251–277.

Whyte, Kyle Powys. "Our Ancestors' Dystopia Now: Indigenous Conservation and the Anthropocene." In *The Routledge Companion to the Environmental Humanities,* edited by Ursula K. Heise, Jon Cristensen and Michelle Niemann, 206–215. London: Routledge, 2016.

Williams, Peter. *Snail.* London: Reaktion Books, 2009.

Wilson, Edward O. "The Little Things That Run the World (the Importance and Conservation of Invertebrates)." *Conservation Biology* 1, no. 4 (1987): 344–346.

Winter, Kawika, Kamanamaikalani Beamer, Mehana Vaughan, Alan Friedlander, Mike Kido, A. Nāmaka Whitehead, Malia Akutagawa, Natalie Kurashima, Matthew Lucas, and Ben Nyberg. "The Moku System: Managing Biocultural Resources for Abundance within Social-Ecological Regions in Hawai'i." *Sustainability* 10, no. 3554 (2018): 1–19.

Wood, Bill. "Hard Choices at Bishop Museum." *Hawai'i Investor,* October (1985): 18–22.

Woodard, Ben. *Slime Dynamics.* Winchester: John Hunt Publishing, 2012.

Worster, Donald. *Nature's Economy: A History of Ecological Ideas.* Cambridge: Cambridge University Press, 1994.

Yeung, Norine W., and Kenneth A. Hayes. "Biodiversity and Extinction of Hawaiian Land Snails: How Many Are Left Now and What Must We Do to Conserve Them? A Reply to Solem (1990)." *Integrative and Comparative Biology* 58, no. 6 (2018): 1157–1169.

Yeung, Norine W., John Slapcinsky, Ellen E. Strong, Jaynee R. Kim, and Kenneth A. Hayes. "Overlooked but Not Forgotten: The First New Extant Species of Hawaiian Land Snail Described in 60 Years, *Auriculella gagneorum* sp. nov. (Achatinellidae, Auriculellinae)." *ZooKeys* 950 (2020): 1–31.

Yusoff, Kathryn. *A Billion Black Anthropocenes or None.* Minneapolis: University of Minnesota Press, 2018.

Zentelis, Rick, and David Lindenmayer. "Bombing for Biodiversity-Enhancing Conservation Values of Military Training Areas." *Conservation Letters* 8, no. 4 (2015): 299–305.

Index

Intergovernmental Science-Policy Platform on Biodiversity and Ecosystem Services (IPBES), 115–116

International Code of Zoological Nomenclature, 109, 128–129

International Union for Conservation of Nature (IUCN), 8–10, 145
Red List, 9, 115–116, 133

Interstate H-3, 81–82

Invasive species, 17, 27, 30–31, 75. *See also* Nonnative plants

Invertebrates
ecological functions of, 12, 132
endangerment of, 9–12, 131–134
popular attitudes toward, 11–12, 133–134, 144
prevalence of, in animal kingdom, 132
study of, 9, 46, 131–133
taxonomy of, 114, 115, 129–130
Umwelt of, 46

Islands, 73–74

IUCN. *See* International Union for Conservation of Nature

Jackson's chameleons, 2
Jasmine, 52, 64
Joint Base Lewis—McChord, Washington, 151

Ka'ena Point, 197–199, 203, 208
Kahanahaiki, 157–160, 163, 170
Kaho'olawe, 105, 162, 166, 168, 173
Kaho'olawe Island Reserve, 168
Kāhuli, 1, 63, 87–89. *See also* Tree snails
"Kāhuli aku" (mele), 107
Kānaka Maoli
colonization's effect on, 85, 91–100
creation stories of, 25
cultural heritage of, 81
diseases' impact on, 66, 93–94
dispossession of, 17, 66, 86, 94–96, 98–99, 109–110
farming by, 52
on Mākua Beach, 158–160

Mākua Valley, 152
relationship to 'āina, 83–84, 86–89, 94–96, 104–108, 206
revitalization of culture, 104–108, 206, 234n81
role of story in culture of, 18
snails' role in lives and culture of, 13, 14, 17, 62–63, 106–108, 208

Kajihiro, Kyle, 161–162, 172–173
Kalākaua, King, 88
Kalo (taro), 3, 89, 103–104
Kamehameha I, King, 65–66, 91, 227n28
Kamehameha III, King, 94
Kanahele, Pualani Kanaka'ole, 84
Kanaka'ole, Edith, 76
Kauanui, J. Kēhaulani, 95
Kaona (multiplicity of meanings), 77
Kapu system, 91
Kaua'i, 6, 60, 62, 66, 93
Kawamoto, Regie, 121–123
Kay, E. Alison, 92, 121
Keiki (young snails), 21, 36–37, 177, 180–181, 183, 184
Kimura, Larry Lindsey, 108–110
Kino lau (god forms), 84, 88, 107, 230n9
Kīpuka (forest areas preserved from lava flows), 106, 178, 182
Knowledge, 76–78
Kondo, Yoshio, 58, 117–120, 123, 205
Ko'olau Range, 19, 21, 51, 102, 171, 177, 191, 201
Kuahu (altars), 83–84. *See also* Ahu
Kumu hula, 13, 76, 83. *See also* Hula
Kumulipo, 88

Lab. *See* Snail Lab
Laminella aspera, 5
Laminella sanguinea, 4–5, 8, 47
Lāna'i, 2
Land shell fever, 92–93, 110
Language, 98
Laysan albatross, 203
Leopold, Aldo, 69

predatory danger posed by, 2, 3, 27, 30–
32, 52, 156, 171–172, 179, 186, 192,
203, 204
protections against, 26–29
reproductive capacity of, 29
two or more species classified as, 31
Rundell, Rebecca, 202
Rururu, 138

Saltwater, 53, 55
Sandalwood, 17, 65
Sandwich Island Institute, 92
Sandwich Islands, 91, 98
Sato, Aimee You, 62–63, 89
Schofield Barracks, 163
Seabirds, 203
Seawater, 53, 55
Self-fertilization, 59
Sentinel species, 201–202
Sepkoski, David, 101
SEPP. *See* Snail Extinction Prevention Program
September 11, 2001, attacks, 165
Shank, Barbara, 154
Shells
appearance of, 24
growth of, 140
information contained in, 140–141
survival of, 141
thickness of, 140
Sherwood, Michael, 156–157
Sierra Club Legal Defense Fund, 156–157,
159–160. *See also* Earthjustice
Singing, of snails, 13, 61–63, 76, 88, 99, 209
Single island endemics, 71
Sischo, Dave, 1–3, 5, 11, 21, 23–25, 28–29,
33, 42–47, 124, 143, 145, 174–175,
177, 179–180, 184–187, 190–192,
202–203, 205–206
Slime
adult vs. keiki, 37
attitudes toward, 25
communicative functions, 36–40, 45
composition of, 34

DNA collection from, 126
generative nature, 25, 88
homing and, 32, 38–39
locomotive role, 26, 34–35
moisture-preserving properties, 32, 34,
55, 183
origins and functions, 34
perceptive capabilities, 33
significance of, 24
studies of, 36–40
Umwelt (world of meaning) constructed by,
24, 35–36, 41
Slow violence, 64
Smallpox, 93–94
Smith, Linda Tuhiwai, 233n63
Snail Extinction Prevention Program (SEPP),
2–3, 20, 21, 23, 28, 124, 131, 142–143,
172, 177, 179, 190, 202, 205
Snail Lab, 177–195, *181*
behaviors of snails raised in, 43–47
care routine, 180–185
description of, 179–180
expansion of, 205
maintenance of snails in, 44, 180–185
name of, 178
nature and purpose of, 188–189
protected snails housed in, 3, 19, 23,
142–143
risks, 185
site of hope and bearing witness, 178, 182–
183, 186, 193–195
Snails. *See also* Conservation; Extinction
consumption of, 88
ecological role of, 68–70, 72, 207
emergencies, 184–185
energy expenditure of, 140
folk beliefs about, 13, 62–63, 76
habitats of, 34
Hawaiian names for, 89–90
hermaphroditic nature of, 21, 29, 37, 59
nocturnal nature of, 1, 33, 38, 52, 57
popular opinion of, 25
significance of, 78–79, 207–208

environmental destruction linked to, 148–
153, 156–157, 159, 167–168, 190–191
legally mandated efforts in snail conservation, 2, 28, 151, 163, 165–166, 168,
170–172, 176, 190
relations with, 149–150, 161–162, 168
World War II, 148
US Department of Agriculture, 30
US Department of Defense (DoD), 150–
152, 168
US Fish and Wildlife Service (USFWS), 2,
28, 82, 156, 163–164, 168, 174
US Marines, 96
US military. *See also* US Army; US Marines;
US Navy
acquisition of new land, 169
conservation activities, 151–152
conversion of former facilities, 168–169
endangered species on land managed by,
151
global reach of, 153
highway construction by, 81
impact of, on snails and their habitats, 17
US Navy, 162, 168, 173–175

Vancouver, George, 66, 227n28
Varex, 140
Vaughan, Mehana Blaich, 62–63, 206
Vie, Jean Christophe, 145
Vitrina pellucida, 58

Wai, Leandra, 158–159, 161, 166, 173
Waiʻanae Range, 200–201
Wallace, Alfred Russel, 55, 235n85
Wandering, 42–47, 49
Welch, d'Alté A., 154
Williams, Peter, 24
Wilson, E. O., 12
Wind, snails blown on, 58–59
Winter, Kawika, 4, 65, 206
Witnessing, 193–195
Wolfsnails. *See* Rosy wolfsnails

World War II, 148
Wright, Ann, 174

Yeung, Nori, 5–6, 10–11, 58, 110–114, 116,
118, 123–126, 128, 131, 133, 135–137,
164–165, 192, 204–205
Yew, Joanne, 36
Yusoff, Kathryn, 221n28

Zoological Society of London, 27